THE MILITARY
MAXIMS OF
NAPOLEON

THE MILITARY MAXIMS OF NAPOLEON

Translated from the French
by
Lieutenant-General Sir George C. D'Aguilar,
C.B.

With a new Introduction and Commentary
by
David G. Chandler

Greenhill Books

This edition of *The Military Maxims of Napoleon*
first published 1987 by Greenhill Books, Lionel Leventhal Limited,
Park House, 1 Russell Gardens, London NW11 9NN

This edition © Lionel Leventhal Limited, 1987
New Introduction and Commentary © David G. Chandler, 1987

British Library Cataloguing in Publication Data
Napoléon *I, Emperor of the French*
The military maxims of Napoleon. – New 1987 ed.
1. Military art and science
I. Title   II. D'Aguilar, *Sir*, G.C.
III. Napoléon *I, Emperor of the French*
Officer's Manual
355'.02'01   U102

ISBN 0-947898-64-6

Publishing History
This new edition of *The Military Maxims of Napoleon* is
based upon Lieutenant-General Sir George C. D'Aguilar's translation
which was first published in 1831 and reissued with an Introduction
by William E. Cairnes in 1901 (Freemantle & Co., London).
For this new 1987 edition David G. Chandler has
contributed a new Introduction and Commentary.

Printed and bound in Great Britain by The Bath Press,
Lower Bristol Road, Bath.

CONTENTS

TABLE OF MAXIMS

To

Those Officers of the British Army

*into whose hands
the original may not have fallen, or who
may have wanted an opportunity
of combining the study of
science with its practical application, this little
volume is inscribed as a fresh incitement to
professional enterprise and acquire-
ment by their faithful and
obedient servant*

THE TRANSLATOR

GENERAL INTRODUCTION
by David G. Chandler

NO major edition of Napoleon's *Military Maxims* has been published in Great Britain since the turn of this century, notwithstanding there having been many, many dramatic military events upon which comment in the context of Napoleon's military genius might well be made. Consequently it would seem an apposite moment to produce an up-dated version of the 1901 edition, retaining the observations of William E. Cairnes relating to the Great Boer War and the original annotations dating back to 1831 or possibly earlier, but adding, in the main, comments on the basis of selected dramatic military happenings of this century – the two World Wars, Korea, the post-1945 struggles associated with decolonisation, Vietnam, the Falklands, Grenada, and the Middle East's whole range of conflicts from 1948 to the present, not to exclude the Gulf War and current events in Afghanistan – both of which wars have now long passed their seventh anniversaries. Napoleon would today have been more of a "hawk" than a "dove of peace", and his views on the on-going nuclear debate would have been direct and not to the liking of the faint-hearted. "It is a principle of war", he once wrote, "that when it is possible to employ thunderbolts their use should be preferred to that of cannon."

What, then, is a "maxim"? *The Concise Oxford Dictionary* defines the term as coming from the Latin adjective *maxima*, "greatest": "A general truth drawn from science or experience; principle, rule of conduct." Given Napoleon's proclivity for rattling off ideas in rapid succession (he could, in the pre-shorthand era, keep five secretaries simultaneously employed in writing different letters to his dictation, pacing from one to another as they sat around his office to give each a sentence at a time without ever losing or confusing his separate chains of thought), we might be excused for wondering if there was any possible connection with Sir Hiram S. Maxim, inventor of the single-barrelled quick-firing machine-gun that came to bear his name.

The practical value of military maxims can be debatable. Napoleon, of course, never formulated a precise system of warfare on paper. His genius was essentially that of a practical soldier-statesman, rather than a theorist *per se*, and he also saw no point in giving his concepts to even the marshalate, never mind his opponents or posterity. Consequently the collecting of his *obiter dicta* into any kind of military rule-book for future generations to apply is a process fraught with perils and pitfalls. When writing to subordinates, Napoleon invariably tailored his orders or advice to what he wished the particular recipient at the particular time to receive in order to deal with a specific situation, actual or impending. Consequently it is possible to find many instances in the massive *Correspondance* of apparently contradictory advice being offered in different places – just as it is all too easy to find at first sight contradictory passages in

Holy Writ justifying or challenging almost every conceivable type of human conduct – particularly if the Old Testament is taken too literally into account. Thus great care has to be taken in reading any particular message of generalised significance into any particular maxim; much lies in taking into account the precise circumstances surrounding a Napoleonic pronouncement before assessing its timeless (or otherwise) validity.

Napoleon was by birth a Corsican brought up as a boy to speak Italian and, despite his immense intellect, there is evidence that he never to the end of his life fully mastered the intricacies and finer points of the French language, particularly its subtle use of idiom. Consequently small misunderstandings could creep into his expressions whether oral or written – a factor that has only served to increase the bafflement of later generations of military students seeking the message of the master through the medium of his utterances or writings. For example, one of his best-known sayings is the following: "The principles of war are the same as those of a siege. Fire must be concentrated *on a single point* [author's emphasis] and as soon as the breach is made the equilibrium is broken and the rest is nothing." As the late Sir Basil Liddell-Hart pointed out, most commentators have seized upon the "single point" aspect and have ignored the really crucial word "equilibrium", which indubitably holds the real point the Emperor was trying to convey: it is through upsetting the foe's psychological "balance" that the road to success lies. Further controversy has been attracted to the word "point", some maintaining that Napoleon was implying the strongest sector of an enemy's defences, others that he meant the weakest. However, the actual study of the campaign that gave rise to the comment in the first instance, the operations fought against Piedmont in 1794, makes it highly probable that Napoleon really meant the "hinge" or "joint" of the enemy dispositions. Thus a small matter like the careless use of a single word can lead to doubt and misrepresentation. Many ideas have been attributed to Napoleon in works such as the *Maxims* that he would be the first to refute if his intention was to illustrate a specific point rather than propound a general principle.

Nevertheless, despite such "minefields" awaiting the unwary, and the immense changes that have affected the conduct of warfare since 1815 – particularly at the technological level – there is much still to be found of relevance and value in pondering Napoleon's *Military Maxims*, providing all necessary allowance is made for those changes and possible inner meanings. All wars, regardless of period, are more like one another than any other form of human undertaking, as Professor Michael Howard has sagely remarked. Any campaign – taken properly in context – can be studied to considerable effect, if only to indicate what *not* to do in certain situations, "and that", as the Duke of Wellington remarked of the Grand Old Duke of York's Flanders Campaign of 1794 (which inspired the famous nursery rhyme), "is always something". The fact is that the basic essentials of warfare and leadership have barely changed over the ages of recorded history despite many superficial appearances to the contrary. And, as will be seen in the new annotations being added to those originally written in

1831, many (although not quite all) of Napoleon's maxims still have relevance when applied to the most recent struggles of the strife-torn 20th century.

A word can usefully be directed at this point to how the *Maxims* are organised and the sequence in which they fall. There is no regular pattern discernible, although there are one or two groupings. Whoever culled the selection from the Master's writings or *obiter dicta* did it in a fairly haphazard fashion – which once and for all rules out Napoleon as the true selector of his own *Maxims* (he was a masterly proponent of logical thought) and also rules out all but the civilian members or generals of his household at Longwood on St Helena. His secretary, Las Cases, fits the first case, but only for the first year; the loyal but rather dim General Bertrand, "Grand Marshal of the Palace", the second, as does the able General Gourgauld and the somewhat sinister Comte Montholon, both of whom are also known to have kept private notes of their master's sayings and taken down extracts from the memoirs he dictated to them, but it is unlikely that soldiers of experience would have set these down in so haphazard a fashion. Another possibility, to which we will revert a little later, is that they may have been the work of a third party, a French compiler (professional or amateur) who never visited St Helena but had access to various papers emanating from the island and to Napoleon's huge documentation.

To make any sense of the 78 maxims, therefore, the reader needs to study the Table of Maxims with some care. For his further assistance, a generalised breakdown by subject areas has been added as Appendix B.

Although all the *Maxims* have some bearing, directly or indirectly, on generalship and the exercise of command in war, their random study can cause some little confusion to the unwary.

Once again, it must be stressed that the *Military Maxims* only provide at best a most generalised guide to actual military conduct. A slavish adherence to the letter can lead to disaster. This has perhaps never been illustrated better than in the events of the first French offensives in Champagne and into Alsace and Lorraine in the autumn of 1914, where General Joffre's blinkered belief – culled from his reading of Napoleon's campaigns of a century earlier – that the remorseless application of the offensive was the only way to a quick victory led to tremendous casualties for minimal gains. In some early attacks, it is alleged, certain French units actually advanced against German machine-guns and barbed wire with uncased *tricolores*, their officers carrying drawn swords in white-gloved hands and the *poilus* dressed in *képis*, sky-blue coats and even scarlet trousers. Military traditions invariably die hard, and probably rightly so in most instances, but there are clear limits to the lengths that *élan* should be taken. No circumstance in war ever reproduces another precisely – each event is unique in one, several or many ways. But those who claim, for instance, that Napoleon has nothing to say on Revolutionary Guerrilla Warfare should ponder the validity of Napoleon's definition of the type of warfare that faced his marshals in Spain – "a war without a front". What could be more concise or accurate than that, at least at the military operational level?

To demonstrate the latent perils inherent in too direct or literal application of the *Maxims* to modern warfare, any attempt to apply the Napoleonic dictum "march dispersed, fight concentrated" on a future European Central Front battlefield could well lead to the presentation to the enemy of a very tempting target for a tactical short-range nuclear weapon (should these continue to form part of modern armies' arsenals, as seems very likely). Modern requirements are therefore for formations to fight in formations as dispersed as is consistent with achieving their appointed roles, relying on artillery, minefields, heli-borne firepower (especially against armour) and conventional airstrikes to reduce the enemy's striking force and, above all, the exploitation capacity of Echelon B reserves. It has been calculated that a modern company commander has at his disposal, or can call down, as much weight of firepower as a Napoleonic commander of a *corps d'armée*. The old saying about every soldier having a "marshal's baton in his knapsack" has thus taken on an unexpected modern meaning. Napoleon, of course, had no air element in his armoury – and indeed may be said to have compromised his "C3I" capacity (Command, Communication and Control, with Intelligence) by ordering the disbandment of the French Army's single *aérostatier* unit of Montgolfier balloonists soon after becoming First Consul. This ill-judged decision – which well illustrates that there were distinct limits to even Napoleon's powers of intellectual vision – is hard to comprehend. The unit had played an important role in the battle of Fleurs (1793), despite the fact that the Austrians captured one balloon half inflated in a church. (It is still to be seen in the National Military History Museum in Vienna.) Had Napoleon possessed such an instrument of intelligence gathering (the equivalent, perhaps, to modern satellite surveillance), the approach of Blucher's three Prussian corps through the Bois de Paris on 18 June 1815 would have been detectable from at least 10 a.m. instead of 1.30 p.m., with possibly incalculable results on the outcome of the climacteric Battle of Waterloo. On the other hand, the same maxim under consideration in this paragraph has indubitable validity for both conventional and guerrilla-type "hit-and-run" wars. The employment of *Military Maxims* as a means of sharpening one's wits by offering points of comparative reference when studying more recent events or speculating on the future is clearly the correct course to pursue. That this process knows few limits of country, political organisation or clime is illustrated by the modern applications of Napoleonic maxims by the armed forces of the Soviet Union.

Soviet military historians of note working in the Napoleonic era – including A. Z. Manfred and L. A. Zak (both contributors to the *Great Soviet Encyclopaedia*) and Lieutenant-General Pavel A. Jiline (late chief of the Military History Institute of the Soviet Army and author of several distinguished works on the Russian view of the Campaign of 1812 in general, and of the contribution to Russia's victory of Kutusov and Davydov's Russian partisans in particular) – pay considerable attention to Napoleon's methods and maxims, even if their main approach is unsurprisingly devoted to portraying Napoleon as an imperialistic menace who fell victim to the superior generalship of Kutusov and the staunch fighting qualities and firm patriotic resolve of the ordinary Russian people during what is usually called "the First Great Patriotic War". In Volume 5 of the *Encyclopaedia* the writer grants that "the French military leader Napoleon I made a significant contribution to the theory and practice of the art of war. He gave divisions and corps a more orderly organisation and cut back sharply on transports, as a result of which the army acquired greater mobility. Napoleon I established the utter defeat of enemy personnel in one all-out battle as the primary goal of combat action and he strove constantly to destroy the enemy by parts, achieving maximum superiority of forces on the axis of the main effort. Very significant in the development of Russian military science", this source inevitably continues, "was the leadership skill of M. I. Kutusov, who was able to crush Napoleon's army, which was one of the leading armies at the beginning of the 19th century." In Volume 15 under "War, Art of", the author closely links the two figures: "The French military commander Napoleon I and the Russian general M. I. Kutusov played an important role in improving the new methods of carrying on military operations in the late 18th and early 19th centuries. Napoleon I introduced the massive use of artillery and cavalry and employed reserves skilfully to achieve a turning point in the battle. Kutusov achieved victories by carrying on a series of successively small and large battles." The Emperor does not get off scot-free, however. In Volume 24 under "Military Strategy", we find that "Napoleon discovered the correct strategic use of massive armed forces for defeating enemy forces, but his military strategy, expressed in a desire to conquer the world, was on the whole adventuristic".

The Soviet view of the modern value of maxims or "abstract truths" – even when emanating from the great Kutusov – is expressed as follows by General Jiline: "But, of course, there is no such thing as an abstract truth which can apply for all times. Truth is always specific. V. I. Lenin called it a 'useless venture' to employ old combat methods under new conditions and in a different historical situation, although they may have yielded excellent results in their time. One cannot in any way compare the nature of contemporary wars, on their gigantic scale, and with the participation of millions in armies with diverse military equipment, with the wars of the past. The size, the content and the outcome of the operations carried on by the Soviet army in the [Second] Great Patriotic War has no equal in war history. They were prepared and conducted on the completely new bases of Soviet military science." (From P. A. Jiline, *Kutusov: his life and activity as a military leader*, Moscow, 1978, chapter 7, "Military theorist".)

The same historian (whose views as a Corresponding Member of the USSR Academy of Sciences carry considerable weight) gives an interesting evaluation of the place of Napoleon's *Maxims* in 19th century Tsarist Russia. "If M. I. Kutusov's military legacy was depreciated or simply ignored in Russian official historiography of the 19th century, then Napoleonic military art was exalted to excess. Its active propagandist was General Jomini, of Swiss origin, who served in Russia in 1813. In his work on the military art, Napoleon's principles for waging an armed struggle were proclaimed to be perpetual and unalterable. Enthusiasm for Napoleon's military art had a negative effect on the development of Russian military thought and inevitably brought with it a devaluing of Russian military experience and the legacy of the great Russian military leaders, including Kutusov. In 1842, *Rules, thoughts and opinions of Napoleon about military art* gathered by F. Kauzler, was published and widely popularised in the army. The aim of this publication, as the translator Y. Leontev pointed out, was to spread 'the most basic and correct understanding of military matters' among young officers . . . In 1846 was published *Napoleon's Military Rules*, translated from the French by N. P. Yershove. This gave, the translator stated, recipes for all instances of military practice and information as to how "reading these rules of the immortal military leader should be adopted' . . . A year later was published *Rules of Conducting War extracted from the writings of Napoleon, the Archduke Charles, General Jomini and other Military Writers*, compiled by M. I. Bogdanovich, Professor of the Military Academy [presumably at St Petersburg; these arm-chair pundits seem always to have been getting into the act!] However, Pavel Jiline concludes, 'no place is found in it for the military views of Russian military leaders such as Peter I, Rumyantsev, Suvorov and Kutusov'." Ah, well! No commentator can hope to be perfect.

Communist China's views of Napoleon and his *Maxims* are harder to discover. Even allowing for Chinese conviction concerning their effortless historical and intellectual superiority over any number of relatively recent "red barbarians" produced by the so-called Western civilisations from the Pharaohs' times onwards and their faith in the military writings of T'sun Tzu from *c.*500BC allied to those tenets expressed in *The Little Red Book* of a rather more recent Great Leader, there are singularly few Napoleonic references or *obiter dicta* to be found in the voluminous works of Chairman Mao. Perhaps a little more attention, if only in passing, will be forthcoming from the now more open-minded and outward-looking Military Historical Institute of the People's Army, which has very recently made a few welcome (if tentative) approaches to this editor. Whatever exchanges of views these may lead to, there is one Napoleonic saying (not, however, included in the 1901 edition of the *Maxims*) that is clearly of modern relevance: "Let China sleep, for when she wakes the world will rue it." Despite this rather Nostradamus-type or "Old Mother Shipton-esque" statement, it is possible that "the Corsican Ogre" had a point. Clearly, as even the Soviets must agree, Napoleon's "adventuristic imperialism" did not quite extend to the Great Wall of China. He was prepared to settle for the achievement of Alexander the Great, with a subservient Europe and perhaps Russia thrown in. The relaxed Chinese view is probably the modern parallel of the anonymous Islamic historian who simply recorded the dire events of 1798 in Egypt and the Middle East thus: "This year the pilgrimage to Mecca was discontinued." *Sic transit gloria mundi.*

It is now well over a century and a half since Napoleon breathed his last on the South Atlantic island of St

Helena and almost as long since the first publication of quotations in French and English purporting to be the "sheep-worrier of Europe's" (as Carlyle dubbed him in the Victorian age) *Military Maxims*. In fact, it is not always appreciated that the first such document appeared a year before the Emperor's demise. In the Central Library at Sandhurst there is a slim printed volume entitled *A Manuscript found in the portfolio of Las Casas [sic] containing maxims and observations of Napoleon, collected during the last two years of his residence at St. Helena.* (Translated from the French; London; printed for Alexander Black, Foreign Library, 27 Pall Mall, 1820.) Emmanuel August Dieu-Donné Marius Joseph, Marquis de Las Cases, as is well known, had taken much dictation from the ex-Emperor before falling foul of the British Governor, Sir Hudson Lowe. In consequence of this he was arrested, had all his papers seized, and was ultimately deported from St Helena in November 1816. Four years later a selection of extracts culled from the somewhat haphazard collection of "sentences, bon-mots, and maxims, collected by Las Casas [sic], in his daily conversations with the prisoner Napoleon, and committed to paper literally as he heard them", appeared on both sides of the Channel. Soon after 1823 a copy had reached the shelves of the Gentlemen Cadet's Library at the (then) Royal Military College at Sandhurst. What use the young gentlemen aspiring to His Majesty King George IV's commisions in the Land Forces made of this slender 138-page volume with its frontispiece of Wellington (of all people!) comprising a vast store of assorted wisdom on matters military, diplomatic and political, is not certain, but its near-mint condition would suggest very little. This would not have entirely displeased the putative source of the *Maxims and Recollections*. Napoleon had little time for abridgements. "Extracts?", he once declared, "A pitiable method. Young people have time to read long books . . ." Possibly, although their late 20th century counterparts seem to show scant desire, even if they have the time, to drink deeply at the well of historical knowledge.

In any case this compilation was only one of a number to be published in English during the 19th century, and it contained all of 469 Maxims. Professor L. E. Henry's *Napoleon's War Maxims* (London, 1900), dedicated to HRH The Duke of Clarence and Avondale, K.G., Major of the 10th Royal Hussars, "with whom he worked and whose Royal friendship he greatly valued", contains 115 fully annotated maxims (the first 78 of which in Part I are re-translations of those reproduced in our source). To these the editor had added another 72 "Thoughts relative to the art of war" in Part III, and a further 344 "Social and political thoughts" in the fourth and final section of the book – none of which, of course, is dealt with here. However, the 1901 edition of *The Military Maxims of Napoleon*, which forms the basis of our present venture (in 1987), contains just 78 maxims, the same as the first 78 included in the 1900 version but clearly translated from the French, as the title-page declares, by Lieutenant-General Sir George C. D'Aguilar, C.B., who would seem to have originally completed this task by the early 1830s as the British Library catalogue includes a Dublin version dated 1831 translated by "Colonel

D'Aiguilar". A third edition is listed under the year 1861, and there were two American editions of 1862 and 1865 respectively, the first being published in the Confederate capital of Richmond, Virginia, and the second emanating from New York in the year of the Union's triumph, besides another undated Kansas City edition using D'Aguilar's translation with notes by General Burnod. The 1901 London edition does not proceed to give "Book Two", comprising Maxims 79 to 115 in the Henry version. Why it does not do so is uncertain, but the editor of the early 20th century version, William E. Cairnes, would seem to have been content to end with the oft-quoted maxim requiring a would-be Great Captain of the future to read and re-read the campaigns of Napoleon's nap selection of seven great captains of antiquity and the more recent past.

Captain William Elliot Cairnes (1862–1902) was a soldier, inventor and military commentator of some note in late Victorian England. He was commissioned into the Irish militia in 1882, and two years later transferred in the same rank to the post-Cardwell regular army, passing rapidly through two regiments in a six-month period for undisclosed reasons before ending up in the Royal Irish Fusiliers in July 1884. He received his captaincy in 1890, and seven years later became adjutant to the Yorkshire Light Infantry; because of this appointment he did not serve in South Africa. Nevertheless he became a well-known and respected commentator on the Boer War, writing a daily article for the *Westminster Gazette* from 1899 to 1901. In the latter year he served as secretary to a committee set up to enquire into that perennial subject in Great Britain – then as now – the education and training of army officers, and was commended for his work. He was at the same time secretary to the official enquiry into the Remount Department (something of a *cause célèbre* of the day) and also managed to write a stream of seven books during 1900 and 1901. Four were published anonymously: *An absent-minded war* (1900), which gained some popularity and notoriety for its sarcastic tone; *The Army from within* (1901); *A commonsense army* (1901); and *The Military Maxims of Napoleon* (1901). The remainder appeared under his name (*Social Life in the British Army* (1900), *The Coming Waterloo* (1901) – an appeal for the rapid modernisation of the pre-Haldane British Army – and in 1901 again, *Lord Roberts as a Soldier in Peace and War*). This tremendous, almost frenetic, literary output allied to his other duties not wholly unforeseeably brought on an attack of pneumonia and an early death. Clearly William Cairnes was what today would be called a workaholic – but his writings and committee work had earned him commendation before his demise, and he may be credited with being one of the influences for reform, alongside such fellow-authors as Spenser-Wilkinson, who used the hard experiences of the Great Boer War to demand the overdue modernisation of the complacent British Army – carried through in large measure from 1905 by the Secretary of War, Viscount Haldane, just in time for the supreme test of the First World War.

The original translator of Napoleon's *Military Maxims* belonged to an earlier generation of soldier and was also a figure of some distinction in his day. Lieutenant-

General Sir George Charles D'Aguilar (1784–1855), born at Winchester, was commissioned into the 86th Foot in 1799. He was promoted lieutenant in 1802 and for meritorious service had become a brigade-major just four years later. He saw much service in the Mahratta Wars, being severely wounded at the siege of Bhurtpore. In later years he served as a member of military missions sent to Yanine and Constantinople. He just missed fighting at Waterloo, but in 1817 was gazetted major in the Rifle Brigade. A series of appointments led to eventual promotion to major-general (1841) and then to a command in the First China War, which came to a climax with the capture of Canton in 1847. Ten years later he was promoted lieutenant-general and made CB. His translation of Napoleon's *Military Maxims* was first published in 1831 and by 1888 was in its third edition. The 1901 publication would therefore appear to be the fourth – and this new edition the fifth. General D'Aguilar was also something of an expert on military law, and wrote *Observations on the practice and functions of courts-martial and courts of enquiry* (published in Dublin in 1859 with a second edition appearing in 1867). It is likely that the annotated *Maxims* are translations of those provided by General Burnod for various French editions. Clearly the commentator is a Frenchman, as the Peninsular War is not mentioned once, and Waterloo appears on only a single occasion, in connection with Maxim XLIX.

What more can we say about the provenance of these celebrated *Military Maxims*? As already stated, their background is cloaked in mystery, and we shall probably never know who first selected them – and indeed there were clearly several versions of varying lengths. That at least the majority do emanate from Napoleon's writings or utterances, there is little doubt. Quite a few are to be found in Volume XXXI of *La Correspondance de Napoléon 1er* (Paris, 1870). This volume comprises part of the ex-Emperor's works dictated on St Helena. Included are "Eighteen notes on the work entitled 'Considerations on the art of war'", a publication that appeared in France in 1816 written by Lieutenant-General Baron Rogniat, a sapper. Napoleon had little time for this publication, and his "Notes", dictated, it seems, to General Bertrand at Longwood, contain many corrections and marginal annotations in Napoleon's own hand. Quite a few passages have passed into the *Maxims*, either verbatim or in slightly adjusted format. Others have been culled from the voluminous *Correspondance* itself: various proclamations, orders of the day going back in some cases to as early as 1795, and the pages of the official government publication, *Le Moniteur*, clearly by a person or persons unknown long before the official publication of the 32-volume series at the instigation of Napoleon III in 1870. It would indeed seem that there were "leaks" occurring even before the latter's distinguished uncle had breathed his last in 1821.

Naturally enough there have been many French publications; by 1874 the *Maxims* were in their fifth edition, and a second impression of this was the basis for an edition of 1900, revised and augmented by Generals Burnod and Husson. There had been a Spanish edition, published in Madrid as early as 1821, and a Swedish

translation appeared in Stockholm four years later. The
20th century has seen several more publications, including *Napoleon and modern war: his military maxims*, edited and
annotated by Conrad C. Lanze (Military Service
Publishing Co., Harrisburg, Pennsylvania, 1943). A
Spanish-language edition compiled by General José
Antonio Paez (including both the 78 maxims and all
of 416 *Pensamientos de Napoleon Primero*) appeared in
Venezuela in 1971 to mark the 150th anniversary of the
glorious and never-to-be-forgotten victory of Carabobo
(1821), where the great Simon Bolivar – aided by a
British battalion of volunteers – crushed the Spanish
army of General Miquel de la Torre and thus ensured
both the capture of Caracas five days later and ultimately the freedom of the country from the Spanish
colonial yoke. In 1976 a French-Canadian version of
Honoré de Balzac's prefaced 1838 edition of *Maximes et
Pensées de Napoléon recueillies par J. L. Gaudy (Jeune)*
appeared with an introduction by Dr Jan Doat of the
Academy of Letters and Human Sciences of the Royal
Society of Canada with a preface by the well-known
Napoleonic specialist Ben Weider, c.m., whose later
books on Napoleon's supposed murder on St Helena
appeared in 1978 and 1982 written in conjunction with
the Swede Sten Forshufvud and with David Hapgood
respectively. This 1976 version contained 525 maxims
and thoughts organised under four main headings (*Des
Hommes, La Guerre* – comprising some 74 entries, but
many being different from both the Las Cases and the
D'Aguilar versions – *Le Monde*, and *Napoléon par Lui-
Même*), and ending with Napoleon's diatribe *Sur*

l'Angleterre for good measure. Doubtless there have been other editions – as this has been a popular work – but none appear to have been published in England for some little time. A particularly useful work – although not a version of the *Maxims* as such – is C. J. Herold's *The Mind of Napoleon* (New York, 1955).

The modern editor of the Canadian version already referred to above is quite convinced that the great French romantic novelist, Honoré de Balzac, played a major part in selecting and classifying the *Maximes et Pensées de Napoléon*. In his introduction Dr Jan Doat describes how much inspiration Balzac drew from the Napoleonic saga as anyone acquainted with his writings will be aware. He claims that Balzac performed his selecting and editing work over a period of seven years. In 1838, fifteen years after the first appearance of the far less impressive Las Cases version, financial distress forced Balzac to sell his work on this subject to the publisher A. Barbier for 4,000 francs, and the book duly appeared that year with a short preface from Balzac's pen. Many later versions in many languages drew heavily from the Balzac version, although there are substantial differences between it and the work Colonel D'Aguilar translated while he was serving as Deputy-Adjutant-General in Ireland and published in an annotated version at Dublin all of seven years before the Balzacian edition of 1838. Dr Doat quotes to good effect a short passage from Las Cases's *Mémorial de Sainte Helene*, (9 volumes, 1890), which is worth citing as a tailpiece. "Returning from his walk, the Emperor studied a large basket filled with broken silver-ware, which was going to be sent to the town next day.

This formed the indispensable means of supplying our food for a month, thanks to the latest (financial) reductions on the part of the Governor [Lieutenant-General Sir Hudson Lowe]."

The *Military Maxims*, therefore, may be said to represent the brilliant fragments of Napoleon's military thought, some of them dazzlingly deceptive, it is true, for a number are perhaps made up of paste costume-jewellery, slipped in by another hand; but the overwhelming majority are of the greatest value as food for mature thought and reflection. As Balzac claimed of his selection, "the soul of the Emperor passes before us", although we may question his later assertion that "Wellington was an accident" and that "France may say with pride that from the depths of his tomb Napoleon still combats England"! But, as Napoleon apparently said (according to Las Cases's *Maxim* CCCCLXV in his 1820 publication), "that book would be a curious one which contained no falsehood". On that note, this latest editor of the Emperor's *Maxims* rests his case. "War", as Napoleon sagely remarked, is "an immense art which comprises all others." It is also, as he again avowed in 1796, "like politics, a matter of tact".

INTRODUCTION
TO THE 1901 EDITION
by William E. Cairnes

S O long as war endures upon this earth, so long as nations devote annually a certain proportion of their populations, and a considerable portion of their wealth, to the maintenance of their armies, so long as the great commander is held in honour, and communities vie with each other in doing homage to the man who has shown in the field that he possesses the highest attributes of military genius, so long will the maxims of the First Napoleon, that unrivalled master of the art of war, be regarded as precepts to be followed with care and rarely to be neglected with impunity.

It is true that the science of warfare has made giant strides since the days of Napoleon; that the invention first of the breech-loader and subsequently of the magazine rifle, of smokeless powder, of quick-firing cannon capable of tremendous destructive effect at enormous ranges, have revolutionised tactics; but the great principles which should guide the commander in the direction of his armies and the preparation of his plans remain unchanged, as they have remained unchanged since the days of Hannibal and Cæsar. As we read through Napoleon's Maxims we shall find it easy to separate those which deal with the tactics of the battlefield from those relating to the conduct of a campaign.

The former are now of little value or importance; the latter are pregnant with instruction, and carry to this day a lesson which cannot be neglected with impunity by any but the soldier of genius who is able to rise superior to all rule, and, with confident yet discreet audacity, to grasp a victory where the humbler soldier would look for nothing but disaster. Of such a mettle was Napoleon himself, who frequently violated his own maxims, sometimes with success, sometimes with disastrous results; and Stonewall Jackson, who, on more than one occasion when a pedant would have been defeated in strict conformity with the rules of war, acted in defiance of every precept, and snatched a victory from a situation full of peril.

But it is not given to every one to gauge a situation as rapidly and as accurately as Stonewall Jackson, or to conceive the daring operations by which Napoleon himself shattered so many hostile armies; it will be usually advisable for the general of less commanding ability to adhere to the maxims of war laid down by Napoleon, and never to depart from them unless he can demonstrate to his own satisfaction that he is fully justified in so doing.

From every campaign we can draw illustrations of the unfortunate results which are likely to follow the neglect of the commonsense rules prescribed by Napoleon. Few campaigns have been more fruitful of such lessons than the campaign which still drags on its melancholy length in South Africa; when the history of this war is finally written, as it cannot be written for many years to come, we shall find in that history, inscribed in letters of blood, a solemn warning to the future generations of military commanders, who will be able to learn from a study of the failures and successes of the war, that a nemesis dogs the steps of the general who, confident in his own conceit and neglectful of the experience of the great captains of the past, has endeavoured to march to victory as to a ceremonial, fancying that his enemy would melt away at his touch.

Where caution has been happily united to audacity, and both have been joined to a shrewd common sense, we have seen that victory awaits the skilled commander; where ordinary precautions have been neglected, where the prowess of the foe has been underrated and despised, where no attempt has been made to forecast his movements or to deceive him as to our intentions, there disaster has followed, as surely and inevitably as the night the day.

No one was ever more conscious of the influence of chance on the fortunes of a battle than Napoleon; no one ever left so little to chance. No one excelled him in the audacity of his conceptions or the rapidity with which he carried them into execution; no one paid more attention to the most minute details of his military administration, or was more cautious and wary when caution and wariness were prescribed by the situation.

The reader will be able to form a better opinion of the connection between the precepts inculcated in these Maxims, and some of our reverses in South Africa, if a few concrete examples are placed before him.

This can be done in one of two ways; namely, either by taking the Maxims as they come, and by indicating where their violation or neglect has been the cause of reverses, or by taking the misfortunes which we sustained in their proper order, and pointing out to what extent they might have been avoided, had the Maxims of Napoleon been adhered to. Of these two plans the latter appears to me to be the best.

What then was our first "regrettable incident" in this campaign? Surely the subdivision of Sir George White's army in Natal and the detaching of General Penn Symonds, against Sir George White's advice but with his consent, to hold Dundee with a small force in a most exposed position. Let us see if Napoleon has got anything to say on the subject of military considerations being made subservient to political exigencies. We shall read in Maxim LXXII the following: – "A general-in-chief has no right to shelter his mistakes in war under cover of his sovereign, or of a minister, when these are both distant from the scene of operation, and must consequently be either ill-informed or wholly ignorant of the actual state of things. Hence it follows, that every general is culpable who undertakes the execution of a plan which he considers faulty. It is his duty to represent his reasons, to insist upon a change of plan, in short to give in his resignation, rather than allow himself to be made the instrument of his army's ruin. Every general-in-chief who fights a battle in consequence of superior orders, with the certainty of losing it, is equally blamable." Napoleon has a good deal more to say on this subject, for which I must refer the curious reader to the Maxim itself, but I have quoted sufficient to show pretty clearly, that he would not have been likely to have approved of Sir George White's action in deferring to Sir Walter Hely-

Hutchinson's arguments in a matter of such vital importance, as the initial disposition of his troops at the commencement of a campaign.

Our next misfortune in the war was the capture at Nicholson's Nek of one and a half battalions of infantry, and a mountain battery, on the same day on which Sir George White fought the indecisive action of Lombard's Kop. None of the Maxims apply very directly to this particular case, unless it be Maxims number XI and XXXIV. In the first of these we are warned that to "act upon lines far removed from each other, and without communications, is to commit a fault which always gives birth to a second. The detached column has only its orders for the first day. Its operations on the following day depend on what may have happened to the main body," which in this case could not be communicated till too late; "let it therefore be held as a principle, that an army should always keep its columns so united, as to prevent the enemy from passing between them with impunity," as the Boers were able to do in this particular case.

Napoleon is very strong on the danger of allowing such an opportunity to an enemy, and recurs to the subject in Maxim XXXIV, in which he informs us that "It should be laid down as a principle never to leave intervals by which the enemy can penetrate between corps formed in order of battle, unless it be to draw him into a snare." It is easier to observe this rule nowadays than it was in Napoleon's time, as the long range of modern firearms enables troops separated by considerable intervals to cover wide tracts with their fire, so denying them to their enemy. For instance, the tactical dispositions for the attack on the Boer position at Elandslaagte, dispositions which necessitated a wide interval between the Devons making the frontal demonstration and the flank attack by the Highlanders, the Manchesters, and the Imperial Light Horse, would certainly have invited counter-attack and disaster in Napoleon's time, though admirably suited to the circumstances of the present moment. But the principle which Napoleon inculcates was adhered to at Elandslaagte, and neglected at Lombard's Kop and Nicholson's Nek, doubtless for what appeared at the time to be perfectly sound reasons, with the results with which we are all familiar.

Coming now to the brilliant story of the defence of Ladysmith, we shall find little which directly applies in the Maxims, though if we read Maxim XLI, we shall find that the Boers followed the principles laid down by the great master of war in preparating, as he directs, both lines of contravallation, to keep in the besieged, and lines of circumvallation at Colenso and elsewhere, to keep out the army charged with the relief of the beleaguered garrison. And this brings us to the subject of the attempts made by Sir Redvers Buller's army to effect this object, attempts which were for long unsuccessful, and some at any rate of which appear to have been made with but little regard to the maxims of war. For instance, the following may be looked upon as having some reference both to the attempt to free the garrison of Ladysmith by operations based on Maritzburg, in preference to adhering to the original plan of campaign with Bloemfontein as our first objective, and to Sir

Redvers Buller's decision to endeavour to force the passage of the Tugela at Colenso. We read in Maxim XVI: "It is an approved maxim in war, never to do what the enemy wishes you to do, for this reason alone, that he desires it. A field of battle, therefore, which he has previously studied and reconnoitred, should be avoided, and double care should be taken where he has had time to fortify or entrench. One consequence deducible from this principle is, never to attack a position in front which you can gain by turning." Many criticisms based upon this reasoning have been already passed upon General Buller's action in transferring the bulk of his army to Natal at all, and also upon his endeavouring to force the passage of the Tugela by a frontal attack. Nor shall we have any difficulty in finding other Maxims which bear upon the same subject. For instance, in Maxim XXXVII we find: "From the moment you are master of a position which commands the opposite bank (of a river), facilities are acquired for effecting the passage of the river; above all, if this position is sufficiently extensive to place upon it artillery in force." Surely there is here a hint of the vital importance of securing Hlangwane Mountain before attempting to force the passage of the Tugela; for on this eminence there was ample room for placing a powerful artillery in position, as we found subsequently, and with it in our hands the whole of the Boer positions near Fort Wyllie, and along the northern bank of the river, became at once untenable. It is clear from the orders for the battle issued by General Clery, presumably with General Buller's approval, that some glimmering of the importance of this

position had penetrated the minds of the British commanders, but they evidently failed to appreciate its vital
importance, or they would hardly have entrusted its
capture to a thousand or so of *mounted men*.

However, the mistake was made, and dearly was it
expiated.

But there are other Maxims which throw a light upon
some of the events of the reverse at Colenso. If we turn to
Maxim LXXIII we read: "The first qualification in a
general-in-chief is a cool head, that is, a head which
receives just impressions, and estimates things and
objects at their real value. He must not allow himself to
be elated by good news or depressed by bad. . . . Some
men . . . raise up a picture in the mind on every slight
occasion, and give to every trivial occurrence a dramatic
interest. But whatever knowledge, or talent, or courage,
or other good qualities such men may possess, nature has
not formed them for the command of armies or the direction of great military operations." Was there any
incident, startling and dramatic in itself, but in reality
without a vital influence on the fortunes of the day,
which may have affected the judgment of Sir Redvers
Buller during the action and induced him to order a
retirement before the battle had been irretrievably lost?
Surely just such an incident is to be found in the circumstances surrounding the loss of Colonel Long's guns, and
the tragic death of so many gallant officers and men,
under the close view of the Commander-in-Chief himself.
It cannot be doubted that this incident powerfully
affected Sir Redvers Buller's judgment, and led him to
order the withdrawal of his amazed troops. So great was

the influence of this tragic incident that nothing was done to recover the guns, though it is the opinion of many officers who were present that there would have been no difficulty in doing so in the hour of darkness between the setting of the sun and the moon's rising. It may be urged in General Buller's defence that it is extremely doubtful whether any good result would have been obtained by continuing an engagement entered upon in haste and under a false impression, and which must if continued have proved costly in life; but there can be no doubt that when the retirement was ordered the troops were far from considering themselves beaten, and were perfectly ready to renew the conflict on the following day. The fact must not then be overlooked that in withdrawing when he did General Buller preserved the *moral* of his army, a result on which he could hardly have counted had he shattered it in fruitless and bloody attempts to storm the impregnable position of the Boers.

To the long and difficult operations which finally culminated in the successful relief of Ladysmith, the success of which was partly due, in the opinion of many, to the victories won by Lord Roberts in the Free State, few of these Maxims can be held to apply, if we except a sentence in Maxim LXXVII, which throws a certain light on the reason of the failure of the operations ending for the time with the evacuation of Spion Kop. The sentence I refer to reads as follows: "Gustavus Adolphus, Turenne, and Frederick, as well as Alexander, Hannibal, and Cæsar, have all acted upon the same principles. These have been: to keep their forces united, to leave no weak part unguarded, to seize with rapidity on important points." It is the opinion of many competent judges that the chief reason of the failure of the attempted movement to turn the right of the Boer position was the want of rapidity with which it was executed. The preliminary movements were admirably conceived and carried out. Nothing could have been better than the swoop on Springfield by our cavalry, the dash by which the passage of the river was secured, and the secrecy with which the whole movement was carried out – so far. But when a curious lethargy seems to have settled upon our leaders. Though it was obvious that time was an element of the most vital importance, as, our movements once disclosed, the Boers were busily engaged before our eyes in entrenching the rocky slopes which we would be compelled to scale to place our outer flank in a position to turn their lines of circumvallation, priceless days were spent in futile reconnaissance and in hesitating movements destined to be without result. When I say futile reconnaissances I must not be taken to underrate the value of reconnaissances, but these were useless, for, when it was finally decided to seize on Spion Kop, it had been left unreconnoitred, and a precious day had to be spent in securing information which ought long previously to have been in our possession. The lamentable result of the enterprise is too notorious to need any allusion here. The whole episode is an object-lesson of the truth of the statement that much more than fertility of resource and sound strategical judgment is required in a leader; he must add to these the power of rapid decision and a concentration of mind which will enable him to see the right course and to carry it out with all the rapidity possible at any cost.

A somewhat similar state of affairs may be found in the abortive attempt on Vaal Krantz. Here again our troops and our leaders were handicapped fatally by ignorance of the *terrain* on which they were to fight, an ignorance for which the authorities cannot be too strongly blamed, as our armies were left unprovided with accurate maps of a colony which had been for upwards of fifty years under our flag.

Coming then to the actual relief of the beleaguered city and the incidents which accompanied that happy termination of a most arduous campaign, we shall find that here again one of the Maxims of the great Napoleon was disregarded with consequences which were, to say the least, regrettable. In Maxim LI we find the following: "It is the business of cavalry to follow up the victory and to prevent the beaten enemy from rallying." Yet when our victorious troops entered Ladysmith no attempt whatever, except of the most feeble and hesitating description, was made to harass his retreat and capture his train. Lord Roberts has made it plain to us in his despatches that he was disinclined to accept the statement put forward by Sir Redvers Buller, a statement which clashed with the report of Sir George White, to the effect that the enemy had succeeded in removing the whole of their guns, baggage, and camp equipment on the day before the relieving army entered Ladysmith. The Commander-in-Chief, well qualified to express an opinion on the course adopted by General Buller, is evidently under the impression that an energetic pursuit by the mounted troops of the main army, troops whose horses were in good condition, would have been attended by important results. But General Buller was disinclined to run any risks. Ladysmith had been relieved, after many months of arduous campaigning, and the general was averse to entangling his mounted troops in the broken country of the Biggarsberg. He had no doubt vivid recollections of the disaster to the squadrons of the 18th Hussars after Glencoe, and was anxious to avoid any possibility of a similar regrettable incident. Hence his attempt to pursue was but half-hearted, and our success was shorn of the results which might have followed it. No great victory can be won without incurring certain risks. The commander who hesitates to run these risks on occasion will rarely achieve any striking victory, though he may equally avoid any irremediable disaster.

If we now turn to the other side of the theatre of war, where also we had to sustain reverses and to meet with disappointments, we shall find that these reverses were invariably due to a neglect of the precepts which ought to guide a commander in making his dispositions. At the battle of the Modder, where considerable losses were sustained to achieve a result which might have been attained at a comparatively small cost, the British commander acted on his own preconceived notion of the situation without paying due heed to the reports of his cavalry. Had Lord Methuen studied Maxim LXXV he would probably have avoided this error and have approached the river with a wariness and caution which was conspicuously absent. Both after the battle of the Modder and after Magersfontein Lord Methuen was bitterly assailed with criticism. At the present moment no good purpose will be served by reviving this discussion, especially when we recollect the excellent services which that commander has since rendered. His earlier mistakes were due to a want of experience and to errors of judgment which he would probably be the first to admit and deplore. Since then he has gained in experience; his judgment has ripened; and as he has preserved throughout that dogged courage and insistent resolution is a manner from which it is impossible to withhold one's admiration, he has now become a cool, daring, and reliable leader, well fitted to command, and in the end will probably rank as one of the finest figures of the campaign. It is impossible not to admire the fortitude with which he has endured in silence the fierce attacks made upon him by his own countrymen in the earlier stages of the war, or the fine resolution which he has displayed in sticking undauntedly to his post through thick and thin, through good and evil. No critic of Lord Methuen's earlier operations can overlook the gallantry with which he has expiated his faults, or the aptitude he has shown to learn the lessons of the war.

In the central portion of the field of operations our arms met with a reverse at Stormberg, a reverse due very much, in the same manner as Lord Methuen's earlier reverses, to the general having adopted on insufficient grounds an erroneous idea of the state of affairs, and to his having neglected to thoroughly reconnoitre the position which he proposed to attack. No one can accuse

General Gatacre of want of courage or of enterprise; but he neglected those precautions which alone can justify the readiness to assume great responsibility and to incur the gravest risks. In his case also one is reluctant to further criticise a commander who has admitted his errors and expiated his faults, serious as they may have been.

There is one thing certain, namely, that the lessons to be learnt from our early failures were not thrown away on the great commander to whom was entrusted the task of restoring our tottering fortunes.

The momentous events which speedily followed when Lord Roberts put his army in motion against the Boers prove that he had not overlooked the great truth set forth in Maxim XVI, in which we are warned "Never to do what the enemy wishes you to do, for this reason alone, that he desires it." And again, "A field of battle, therefore, which he has studied and reconnoitred should be avoided, and double care should be taken where he has had time to fortify or entrench. One consequence deducible from this principle is, never to attack a position in front which you can gain by turning."

When Lord Roberts arrived in South Africa, he found Cronje at Magersfontein in a strongly entrenched position, a position in course of being daily extended so as to threaten to envelop the lines occupied by Lord Methuen, and barring the way to the beleaguered city of Kimberley; while away in Natal Sir Redvers Buller had been brought to a stand by the fortified ridges of the Drakensberg in northern Natal. Bidding Buller to persevere, Lord Roberts proceeded to assist him indirectly, but as was afterwards proved effectively, by preparing to strike heavily at the Free State capital. His preparations completed, he swiftly assembled a large force at Randam, and moving with a decision and a rapidity which formed a strong contrast to our previous deliberate movements, he swept with his horsemen round the Boer positions and forced a way to Kimberley, while throwing a formidable force of infantry across both the Riet and Modder rivers, thus threatening to cut off Cronje from his capital. This decided movement at once rendered the Magersfontein position untenable. Realising the situation, the startled Cronje hastened to fly from this new foe, a foe whom his instinct at once recognised as formidable in the extreme. The sequel is familiar to all of us. Headed off in his flight to Bloemfontein by the rapid movements of French, Cronje was brought to bay in the river-bed near Paardeberg, and after enduring with heroic fortitude for many days a terrific bombardment, was compelled to yield with all his force. Cronje, like Gatacre, fell a victim to his preconceived notions of the tactics of his enemy, notions for which, in his case, he had a certain amount of justification. Lord Roberts then marched direct on Bloemfontein, ousting his foe from one position after another by the skill with which he disposed of his mounted forces. Arrived at Bloemfontein, the question of transferring the line of communications from the west to the south at once became pressing. Such an operation, especially in the face of a mobile and active enemy, is ever most difficult and delicate. This is what Napoleon has to say on the subject. Maxim XX enunciates that: "It may be laid down as a principle, that the line of operation should not be abandoned; but it is one of the most skilful manœuvres in war to know how to change it, when circumstances authorise or render this necessary. An army which changes skilfully its line of operation, deceives the enemy, who becomes ignorant where to look for its rear, or upon what weak points it is assailable." In Lord Roberts' case the operation was imperatively demanded by circumstances. It was impossible to feed or to supply the army by waggon convoys across the long and exposed road to Modder River Station or Kimberley. It was necessary to secure the railway crossing the Orange River at Norval's Pont, a railway running through a country in the occupation of the enemy. Fortunately that enemy had become alarmed by the events passing in his rear, and, pressed in front by the British under General Clements and General Gatacre, hastened northward without loss of time. Within a few days of our entering Bloemfontein, the delicate operation had been successfully carried out, and the line, so vital to our existence, was in our hands.

Lord Roberts now found himself in the embarrassing position well described by Napoleon in Maxim LXX. There we can read: – "The conduct of a general in a conquered country is full of difficulties. If severe, he irritates and increases the number of his enemies. If lenient, he gives birth to expectations which only render the abuses and vexations inseparable from war the more intolerable. A victorious general must know how to employ severity, justness, and mildness by turns, if he would allay sedition or prevent it." We are, I think, at last beginning to realise the truth of this maxim. We have seen Lord Roberts in turn attacked for the well-timed severity with which he endeavoured to cope with the attempts of the enemy to inflame to renewed action those who had laid down their arms, and for the leniency with which he treated such of our opponents as he believed he might be able to attach to our cause. And Lord Kitchener at the present moment is in very much the same difficult situation.

But, in addition to these difficulties, Lord Roberts found himself confronted by the armed forces of the enemy still in the field. With these he well knew how to deal, and, as soon as his preparations had been completed, he marched swiftly northward, ousting the Boers by well-planned flanking movements from one entrenched position after the other, till at last Pretoria fell and the capital of the Transvaal lay once again under the British flag.

This is hardly the place to deal at any length with the events which followed on the occupation of Pretoria, events which are so recent that they must be fresh in the memories of us all. Let us rather turn to some of the Maxims which we have not yet considered and see if we can extract from them anything which may throw light upon the manner in which we ought to seek a solution for the difficulties hampering our efforts at army reorganisation. Napoleon in Maxim LX gives us his views as to the best solution of the recruiting problem. He tells us that every means must be taken to attach the soldier to his colours, and indicates his belief that this can be best accomplished by showing consideration and respect to the old soldier. "His pay likewise should increase with his length of service. It is the height of injustice to give a veteran no greater advantage than a recruit." In this Maxim Mr. Brodrick* may find advice from a soldier whose opinion is well entitled to respect.

In conclusion, let no one imagine that the truths enunciated in the following Maxims have ceased to be worthy of consideration owing to lapse of time. The soldier can best prepare himself for the active exercise of his profession by the serious and close study of the campaigns conducted by the great masters of the art of war. Very little reflection will suffice to distinguish such of the Maxims as are still applicable from those which, founded on the limited range of weapons in Napoleon's day, have become obsolete. In his own words: "Peruse again and again the campaigns of Alexander, Hannibal, Cæsar, Gustavus Adolphus, Turenne, Eugene, and Frederick. . . . Your own genius will be enlightened and improved by this study, and you will learn to reject all maxims foreign to the principles of these great commanders."

*Secretary of State for War, 1901.

PREFACE
TO THE ORIGINAL EDITION

IN forming a collection of these Maxims which have directed the military operations of the greatest captain in modern times, my object has been to prove useful to such young officers as desire to acquire a knowledge of the art of war by studying the numerous campaigns of Gustavus Adolphus, Turenne, Frederick, and Napoleon.

These great men have all been governed by the same principles, and it is by applying these to the perusal of their respective campaigns that every military man will recognise their wisdom, and make such use of them hereafter as his own particular genius shall point out.

And here, perhaps, my task might have been considered finished; but perceiving how incomplete the collection was alone, I have endeavoured to supply the deficiency by having recourse for further illustration to the "Memoirs" of Montécuculli and the "Instructions" of Frederick to his generals. The analogy of their principles with those of Napoleon has convinced me that the art of war is susceptible of two points of view. One relates entirely to the acquirements and genius of the general, the other to matters of detail.

The first is the same in all ages, and with all nations, whatever be the arms with which they fight. Hence it follows that great commanders have been governed by the same principles in all times.

The matters of detail, on the contrary, are under the control of existing circumstances. They vary with the character of a people and the quality of their arms.

It is with a view to impress the justice of this remark that I have sought for facts in different periods of history to illustrate these Maxims, and to prove that nothing is *problematical* in war, but that failure and success in military operations depend almost always on the natural genius and science of the chief.

MAXIMS
extracted from the writings
and conversations of the
Emperor Napoleon I

I

THE frontiers of states are either large rivers, or chains of mountains, or deserts. Of all these obstacles to the march of an army, the most difficult to overcome is the desert; mountains come next, and large rivers occupy the third place.

II

IN forming the plan of a campaign, it is requisite to foresee everything the enemy may do, and to be prepared with the necessary means to counteract it.

Plans of campaign may be modified *ad infinitum* according to circumstances, the genius of the general, the character of the troops, and the features of the country.

III

AN army which undertakes the conquest of a country has its two wings either resting upon neutral territories, or upon great natural obstacles, such as rivers or chains of mountains. It happens in some cases that only one wing is so supported, and in others that both are exposed.

In the first instance cited, viz., where both wings are protected, a general has only to guard against being penetrated in front. In the second, where one wing only is supported, he should rest upon the supported wing. In the third, where both wings are exposed, he should depend upon a central formation, and never allow the different corps under his command to depart from this; for if it be difficult to contend with the disadvantage of having *two* flanks exposed, the inconvenience is doubled by having *four*, tripled if there be *six*; that is to say, if the army is divided into two or three different corps. In the first instance then, as above quoted, the line of operation may tend indifferently to the right or to the left. In the second, it should be directed towards the wing in support. In the third, it should be perpendicular to the centre of the army's line of march. But in all these cases it is necessary every five or six days to have a strong post, or an entrenched position, upon the line of march, in order to collect stores and provisions, to organise convoys, to form a centre of movement, and establish a point of defence, to shorten the line of operations.

IV

WHEN the conquest of a country is undertaken by two or three armies, which have each their separate line of operation until they arrive at a point fixed upon for their concentration, it should be laid down as a principle, that the junction should never take place near the enemy, because the enemy, in uniting his forces, may not only prevent it, but beat the armies in detail.

V

ALL wars should be governed by certain principles, for every war should have a definite object, and be conducted according to the rules of art. War should only be undertaken with forces proportioned to the obstacles to be overcome.

VI

AT the commencement of a campaign, to advance or not to advance is a matter for grave consideration, but when once the offensive has been assumed, it must be sustained to the last extremity. However skilful the manœuvres, a retreat will always weaken the *moral* of an army, because in losing the chances of success, these last are transferred to the enemy. Besides, retreats cost always more men and *matériel* than the most bloody engagements, with this difference, that in a battle the enemy's loss is nearly equal to your own, whereas in a retreat the loss is on your side only.

VII

AN army should be ready every day, every night, and at all times of the day and night, to oppose all the resistance of which it is capable. With this view, the soldier should be invariably complete in arms and ammunition; the infantry should never be without its artillery, its cavalry, and its generals; and the different divisions of the army should be constantly in a state to support and to be supported.

The troops, whether halted or encamped or on the march, should always be in favourable positions, possessing the essentials required for a field of battle; for example, the flanks should be well covered, and all the artillery so placed as to have free range and to play with the greatest advantage. When an army is in column of march, it should have advanced guards and flanking parties, to examine well the country in front, to the right, and to the left, and always at such distance as to enable the main body to deploy into position.

VIII

A general-in-chief should ask himself frequently in the day, What should I do if the enemy's army appeared now in my front, or on my right, or on my left? If he have any difficulty in answering these questions he is ill posted, and should seek to remedy it.

IX

THE strength of an army, like the power in mechanics, is estimated by multiplying the mass by the rapidity; a rapid march augments the *moral* of an army, and increases all the chances of victory.

X

WHEN an army is inferior in number, inferior in cavalry, and in artillery, it is essential to avoid a general action. The first deficiency should be supplied by rapidity of movement; the want of artillery by the nature of the manœuvres; and the inferiority in cavalry by the choice of positions. In such circumstances the *moral* of the soldier does much.

XI

To act upon lines far removed from each other, and without communications, is to commit a fault which always gives birth to a second.

The detached column has only its orders for the first day. Its operations on the following day depend upon what may have happened to the main body. Thus the column either loses time upon emergency, in waiting for orders, or acts without them and at hazard. Let it therefore be held as a principle, that an army should always keep its columns so united as to prevent the enemy from passing between them with impunity. Whenever, for particular reasons, this principle is departed from, the detached corps should be independent in their operations. They should move towards a point fixed upon for their future junction. They should advance without hesitating, and without waiting for fresh orders, and every previous means should be concerted to prevent their being attacked in detail.

XII

An army ought only to have one line of operation. This should be preserved with care, and never abandoned but in the last extremity.

XIII

THE distances permitted between corps of an army upon the march must be governed by the localities, by circumstances, and by the object in view.

XIV

AMONG mountains, a great number of positions are always to be found very strong in themselves, and which it is dangerous to attack. The character of this mode of warfare consists in occupying camps on the flanks or in the rear of the enemy, leaving him only the alternative of abandoning his position without fighting, to take up another in the rear, or to descend from it in order to attack you. In mountain warfare the assailant has always the disadvantage. Even in offensive warfare in the open field the great secret consists in defensive combats, and in obliging the enemy to attack.

XV

THE first consideration with a general who offers battle should be the glory and honour of his arms; the safety and preservation of his men is only the second; but it is in the enterprise and courage resulting from the former that the latter will most assuredly be found. In a retreat, besides the honour of the army, the loss is often equal to two battles. For this reason we should never despair while brave men are to be found with their colours. It is by this means we obtain victory, and deserve to obtain it.

XVI

It is an approved maxim in war, never to do what the enemy wishes you to do, for this reason alone, that he desires it. A field of battle, therefore, which he has previously studied and reconnoitred, should be avoided, and double care should be taken where he has had time to fortify or entrench. One consequence deducible from this principle is, never to attack a position in front which you can gain by turning.

XVII

In a war of march and manœuvre, if you would avoid a battle with a superior army, it is necessary to entrench every night, and occupy a good defensive position. Those natural positions which are ordinarily met with, are not sufficient to protect an army against superior numbers without recourse to art.

XVIII

A general of ordinary talent occupying a bad position, and surprised by a superior force, seeks his safety in retreat; but a great captain supplies all deficiencies by his courage, and marches boldly to meet the attack. By this means he disconcerts his adversary, and if this last shows any irresolution in his movements, a skilful leader profiting by his indecision may even hope for victory, or at least employ the day in manœuvring – at night he entrenches himself, or falls back to a better position. By this determined conduct he maintains the honour of his arms, the first essential to all military superiority.

XIX

THE transition from the defensive to the offensive is one of the most delicate operations in war.

XX

IT may be laid down as a principle, that the line of operation should not be abandoned; but it is one of the most skilful manœuvres in war to know how to change it, when circumstances authorise or render this necessary. An army which changes skilfully its line of operation deceives the enemy, who becomes ignorant where to look for its rear, or upon what weak points it is assailable.

XXI

WHEN an army carries with it a battering train, or large convoys of sick and wounded, it cannot march by too short a line upon its depôts.

XXII

THE art of encamping in position is the same as taking up the line in order of battle in this position. To this end, the artillery should be advantageously placed, ground should be selected which is not commanded or liable to be turned, and, as far as possible, the guns should cover and command the surrounding country.

XXIII

WHEN you are occupying a position which the enemy threatens to surround, collect all your force immediately, and menace *him* with an offensive movement. By this manœuvre you will prevent him from detaching and annoying your flanks, in case you should judge it necessary to retire.

XXIV

NEVER lose sight of this maxim, that you should establish your cantonments at the most distant and best protected point from the enemy, especially where a surprise is possible. By this means you will have time to unite all your forces before he can attack you.

XXV

WHEN two armies are in order of battle, and one has to retire over a bridge, while the other has the circumference of the circle open, all the advantages are in favour of the latter. It is then a general should show boldness, strike a decided blow, and manœuvre upon the flank of his enemy. The victory is in his hands.

XXVI

IT is contrary to all true principle to make corps which have no communication act separately against a central force whose communications are open.

XXVII

WHEN an army is driven from a first position, the retreating columns should always rally sufficiently in the rear, to prevent any interruption from the enemy. The greatest disaster that can happen is when the columns are attacked in detail, and before their junction.

XXVIII

No force should be detached on the eve of a battle, because affairs may change during the night, either by the retreat of the enemy, or by the arrival of large reinforcements to enable him to resume the offensive, and counteract your previous dispositions.

XXIX

WHEN you have resolved to fight a battle, collect your whole force. Dispense with nothing. A single battalion sometimes decides the day.

XXX

NOTHING is so rash or so contrary to principle, as to make a flank march before an army in position, especially when this army occupies heights at the foot of which you are forced to defile.

XXXI

WHEN you determine to risk a battle, reserve to yourself every possible chance of success, more particularly if you have to deal with an adversary of superior talent, for if you are beaten, even in the midst of your magazines and your communications, woe to the vanquished!

XXXII

THE duty of an advanced guard does not consist in advancing or retiring, but in manœuvring. An advanced guard should be composed of light cavalry, supported by a reserve of heavy, and by battalions of infantry, supported also by artillery. An advanced guard should consist of picked troops, and the general officers; officers and men should be selected for their respective capabilities and knowledge. A corps deficient in instruction is only an embarrassment to an advanced guard.

XXXIII

IT is contrary to all the usages of war to allow parks or batteries of artillery to enter a defile, unless you hold the other extremity. In case of retreat the guns will embarrass your movements and be lost. They should be left in position under a sufficient escort until you are master of the opening.

XXXIV

IT should be laid down as a principle never to leave intervals by which the enemy can penetrate between corps formed in order of battle, unless it be to draw him into a snare.

XXXV

ENCAMPMENTS of the same army should always be formed so as to protect each other.

XXXVI

WHEN the enemy's army is covered by a river, upon which he holds several *têtes de pont*, do not attack in front. This would divide your force and expose you to be turned. Approach the river in echelon of columns, in such a manner that the leading column shall be the only one the enemy can attack, without offering you his flank. In the meantime let your light troops occupy the bank, and when you have decided on the point of passage, rush upon it and fling across your bridge. Observe that the point of passage should be always at a distance from the leading echelon, in order to deceive the enemy.

XXXVII

FROM the moment you are master of a position which commands the opposite bank, facilities are acquired for effecting the passage of the river; above all, if this position is sufficiently extensive to place upon it artillery in force. This advantage is diminished if the river is more than three hundred toises (or six hundred yards) in breadth, because the distance being out of the range of grape, it is easy for the troops which defend the passage to line the bank and get under cover. Hence it follows that if the grenadiers, ordered to pass the river for the protection of the bridge, should reach the other side, they would be destroyed by the fire of the enemy; because his batteries, placed at the distance of two hundred toises from the landing, are capable of a most destructive effect, although removed above five hundred toises from the batteries of the crossing force. Thus the advantage of the artillery would be exclusively his. For the same reason, the passage is impracticable, unless you succeed in surprising the enemy, and are protected by an intermediate island, or unless you are able to take advantage of an angle in the river, to establish a cross fire upon his works. In this case the island or angle forms a natural *tête de pont*, and gives the advantage in artillery to the attacking army. When a river is less than sixty toises (or one hundred and twenty yards) in breadth, and you have a post upon the other side, the troops which are thrown across derive such advantages from the protection of your artillery, that, however small the angle may be, it is impossible for the enemy to prevent the establishment of a bridge. In this case, the most skilful generals, when they have discovered the project of their adversary, and brought their own army to the point of crossing, usually content themselves with opposing the passage of the bridge, by forming a semicircle round its extremity as round the opening of a defile, and removing to the distance of three or four hundred toises from the fire of the opposite side.

XXXVIII

It is difficult to prevent an enemy, supplied with pontoons, from crossing a river. When the object of an army which defends the passage is to cover a siege, the moment the general has ascertained his inability to oppose the passage, he should take measures to arrive before the enemy at an intermediate position between the river he defends and the place he desires to cover.

XXXIX

In the campaign of 1645 Turenne was attacked with his army before Philipsburg by a very superior force. There was no bridge here over the Rhine, but he took advantage of the ground between the river and the place to establish his camp. This should serve as a lesson to engineer officers, not merely in the construction of fortresses but of *têtes de pont*. A space should always be left between the fortress and the river, where an army may form and rally without being obliged to throw itself into the place, and thereby compromise its security. An army retiring upon Mayence before a pursuing enemy is necessarily compromised; for this reason, because it requires more than a day to pass the bridge, and because the lines of Cassel are too confined to admit an army to remain there without being blocked up. Two hundred toises should have been left between that place and the Rhine. It is essential that all *têtes de pont* before great rivers should be constructed upon this principle otherwise they will prove a very inefficient assistance to protect the passage of a retreating army. *Têtes de pont*, as laid down in our schools, are of use only for small rivers, the passage of which is comparatively short.

XL

Fortresses are equally useful in offensive and defensive warfare. It is true they will not in themselves arrest an army, but they are an excellent means of retarding, embarrassing, weakening, and annoying a victorious enemy.

XLI

THERE are only two ways of ensuring the success of a siege. The first, to begin by beating the enemy's army employed to cover the place, forcing it out of the field, and throwing its remains beyond some great natural obstacle, such as a chain of mountains or large river. Having accomplished this object, an army of observation should be placed behind the natural obstacle, until the trenches are finished and the place taken.

But if it be desired to take the place in presence of a relieving army without risking a battle, then the whole material and equipment for a siege are necessary to begin with, together with ammunition and provisions for the presumed period of its duration, and also lines of contravallation and circumvallation, aided by all the localities of heights, woods, marshes, and inundations.

Having no longer occasion to keep up communications with your depôts, it is now only requisite to hold in check the relieving army. For this purpose an army of observation should be formed, whose business it is never to lose sight of the enemy, and which, while it effectually bars all access to the place, has always time enough to arrive upon his flanks or rear in case he should attempt to steal a march.

It is to be remembered, too, that by profiting judiciously by the lines of contravallation, a portion of the besieging army will always be available in giving battle to the approaching enemy.

Upon the same general principle, when a place is to be besieged in presence of an enemy's army, it is necessary to cover the siege by lines of *circumvallation*.

If the besieging force is of numerical strength enough (after leaving a corps before the place four times the amount of the garrison) to cope with the relieving army, it may remove more than one day's march from the place, but if it is inferior in numbers after providing for the siege as above stated, it should remain only a short day from the spot, in order to fall back upon its lines if necessary, or receive succour in case of attack.

If the investing corps and army of observation are only equal when united to the relieving force, the besieging army should remain entire within, or near its lines, and push the works and the siege with the greatest activity.

XLII

FEUQUIÈRES says that we should never wait for the enemy in the lines of circumvallation, but that we should go out and attack him. He is in error. There is no authority in war without exception; and it would be dangerous to proscribe the principle of awaiting the enemy within the lines of circumvallation.

XLIII

THOSE who proscribe lines of circumvallation, and all the assistance which the science of the engineer can afford, deprive themselves gratuitously of an auxiliary, which is never injurious, almost always useful, and often indispensable. It must be admitted at the same time, that the principles of field fortification require improvement.

This important branch of the art of war has made no progress since the time of the ancients. It is even inferior

at this day to what it was two thousand years ago.
Engineer officers should be encouraged in bringing this
branch of their art to perfection, and in placing it on a
level with the rest.

XLIV

If circumstances prevent a sufficient garrison being left
to defend a fortified town which contains a hospital and
magazines, at least every means should be employed to
secure the citadel against a *coup de main*.

XLV

A FORTIFIED place can only protect the garrison and
arrest the enemy for a certain time. When this time has
elapsed and the defences are destroyed, the garrison
should lay down its arms. All civilised nations are agreed
on this point, and there never has been an argument
except with reference to the greater or less degree of
defence which a governor is bound to make before he
capitulates. At the same time there are generals, Villars
among the number, who are of opinion that a governor
should never surrender, but that in the last extremity he
should blow up the fortifications, and take advantage of
the night to cut his way through the besieging army.
Where he is unable to blow up the fortifications he may
always retire, they say, with his garrison and save the
men.

Officers who have adopted this line of conduct have
often brought off three-fourths of their garrison.

XLVI

THE keys of a fortress are well worth the retirement of a garrison, when it is resolved to yield only on those conditions. On this principle it is always wiser to grant an honourable capitulation to a garrison which has made a vigorous resistance than to risk an assault.

XLVII

INFANTRY, cavalry, and artillery are nothing without each other. They should always be so disposed in cantonments as to assist each other in case of surprise.

XLVIII

THE formation of infantry in line should be always in two ranks, because the length of the musket only admits of an effective fire in this formation. The discharge of the third rank is not only uncertain, but frequently dangerous to the ranks in its front. In drawing up infantry in two ranks there should be a supernumerary behind every fourth or fifth file. A reserve should likewise be placed twenty-five paces in rear of each flank.

XLIX

THE practice of mixing small bodies of infantry and cavalry together is a bad one, and attended with many inconveniences. The cavalry loses its power of action. It becomes fettered in all its movements. Its energy is destroyed; even the infantry itself is compromised, for on the first movement of the cavalry it is left without support. The best mode of protecting cavalry is to cover its flank.

L

CHARGES of cavalry are equally useful at the beginning, the middle, and the end of a battle. They should be made always, if possible, on the flanks of the infantry, especially when this last is engaged in front.

LI

IT is the business of cavalry to follow up the victory, and to prevent the beaten army from rallying.

LII

ARTILLERY is more essential to cavalry than to infantry, because cavalry has no fire for its defence, but depends upon the sabre. It is to remedy this deficiency that recourse has been had to horse-artillery. Cavalry, therefore, should never be without cannon, whether when attacking, rallying, or in position.

LIII

IN march or in position the greater part of the artillery should be with the divisions of infantry and cavalry. The rest should be in reserve. Each gun should have with it three hundred rounds, without including the limber. This is about the complement for two battles.

LIV

ARTILLERY should always be placed in the most advantageous positions, and as far in front of the line of cavalry and infantry, without compromising the safety of the guns, as possible.

Field batteries should command the whole country round from the level of the platform. They should on no account be masked on the right and left, but have free range in every direction.

LV

A GENERAL should never put his army into cantonments when he has the means of collecting supplies of forage and provisions, and of thus providing for the wants of the soldier in the field.

LVI

A GOOD general, a well-organised system, good instruction, and severe discipline, aided by effective establishments, will always make good troops, independently of the cause for which they fight.

At the same time, a love of country, a spirit of enthusiasm, and a sense of national honour, will operate upon young soldiers with advantage.

LVII

WHEN a nation is without establishments and a military system, it is very difficult to organise an army.

LVIII

THE first qualification of a soldier is fortitude under fatigue and privation. Courage is only the second; hardship, poverty, and want are the best school for a soldier.

LIX

THERE are five things the soldier should never be without: his firelock, his ammunition, his knapack, his provisions (for at least four days), and his entrenching tool. The knapsack may be reduced to the smallest size possible, but the soldier should always have it with him.

LX

EVERY means should be taken to attach the soldier to his colours. This is best accomplished by showing consideration and respect to the old soldier. His pay likewise should increase with his length of service. It is the height of injustice to give a veteran no greater advantages than a recruit.

LXI

IT is not set speeches at the moment of battle that render soldiers brave. The veteran scarcely listens to them, and the recruit forgets them at the first discharge. If discourses and harangues are useful, it is during the campaign; to do away unfavourable impressions, to correct false reports, to keep alive a proper spirit in the camp, and to furnish materials and amusement for the bivouac. All printed orders of the day should keep in view these objects.

LXII

TENTS are unfavourable to health. The soldier is best when he bivouacs, because he sleeps with his feet to the fire, which speedily dries the ground on which he lies. A few planks and a morsel of straw shelter him from the wind.

On the other hand tents are necessary for the superior officers, who have to write and to consult their maps. Tents should therefore be issued to these, with directions to them never to sleep in a house. Tents are always objects of observation to the enemy's staff. They afford information of your numbers, and the ground you occupy, while an army bivouacking in two or three lines is only distinguishable from afar by the smoke which mingles with the clouds. It is impossible to count the number of the fires.

LXIII

ALL information obtained from prisoners should be received with caution, and estimated at its real value. A soldier seldom sees anything beyond his company; and an officer can afford intelligence of little more than the position and the movements of the division to which his regiment belongs. On this account the general of an army should never depend upon the information derived from prisoners, unless it agrees with the reports received from the advanced guards, in reference to the position, &c., of the enemy.

LXIV

NOTHING is so important in war as an undivided command: for this reason, when war is carried on against a single power, there should be only one army, acting upon one base, and conducted by one chief.

LXV

THE same consequences which have uniformly attended long discussions and councils of war will follow at all times. They will terminate in the adoption of the worst course, which in war is always the most timid, or, if you will, the most prudent. The only true wisdom in a general is determined courage.

LXVI

IN war the general alone can judge of certain arrangements. It depends on him alone to conquer difficulties by his own superior talents and resolution.

LXVII

To authorise generals or other officers to lay down their arms in virtue of a particular capitulation, under any other circumstances than when they are composing the garrison of a fortress, affords a dangerous latitude. It is destructive of all military character in a nation to open such a door to the cowardly, the weak, or even to the misdirected brave. Great extremities require extraordinary resolution. The more obstinate the resistance of an army, the greater the chances of assistance or of success.

How many seeming impossibilities have been accomplished by men whose only resource was death!

LXVIII

THERE is no security for any sovereign, for any nation, or for any general, if officers are permitted to capitulate in the open field, and to lay down their arms in virtue of conditions, favourable to the contracting party, but contrary to the interests of the army at large. To withdraw from danger, and thereby to involve their comrades in greater perils, is the height of cowardice. Such conduct should be proscribed, declared infamous, and made punishable with death. All generals, officers, and soldiers who capitulate in battle to save their own lives, should be decimated.

He who gives the order, and those who obey are alike traitors, and deserve capital punishment.

LXIX

THERE is but one honourable mode of becoming prisoner of war. That is, by being taken separately; by which is meant, being cut off entirely, and when we can no longer make use of our arms. In this case there can be no conditions, for honour can impose none. We yield to an irresistible necessity.

LXX

THE conduct of a general in a conquered country is full of difficulties. If severe, he irritates and increases the number of his enemies. If lenient, he gives birth to expectations which only render the abuses and vexations inseparable from war the more intolerable. A victorious general must know how to employ severity, justice, and mildness by turns, if he would allay sedition, or prevent it.

LXXI

NOTHING can excuse a general who takes advantage of the knowledge acquired in the service of his country to deliver up her frontier and her towns to foreigners. This is a crime reprobated by every principle of religion, morality, and honour.

LXXII

A GENERAL-in-chief has no right to shelter his mistakes in war under cover of his sovereign, or of a minister, when they are both distant from the scene of operation, and must consequently be either ill informed or wholly ignorant of the actual state of things.

Hence it follows that every general is culpable who undertakes the execution of a plan which he considers faulty. It is his duty to represent his reasons, to insist upon a change of plan; in short, to give in his resignation rather than allow himself to become the instrument of his army's ruin. Every general-in-chief who fights a battle in consequence of superior orders, with the certainty of losing it, is equally blamable.

In this last-mentioned case the general ought to refuse obedience, because a blind obedience is due only to a military command given by a superior present on the spot at the moment of action. Being in possession of the real state of things, the superior has it then in his power to afford the necessary explanations to the person who executes his order.

But supposing a general-in-chief to receive a positive order from his sovereign, directing him to fight a battle, with the further injunction to yield to his adversary and allow himself to be defeated – ought he to obey it? No; if the general should be able to comprehend the meaning or utility of such an order, he should execute it, otherwise he should refuse to obey it.

LXXIII

THE first qualification in a general-in-chief is a cool head – that is, a head which receives just impressions, and estimates things and objects at their real value. He must not allow himself to be elated by good news, or depressed by bad.

The impressions he receives, either successively or simultaneously in the course of the day, should be so classed as to take up only the exact place in his mind which they deserve to occupy; since it is upon a just comparison and consideration of the weight due to different impressions that the power of reasoning and of right judgment depends.

Some men are so physically and morally constituted as to see everything through a highly coloured medium. They raise up a picture in the mind on every slight occasion, and give to every trivial occurrence a dramatic interest. But whatever knowledge, or talent, or courage, or other good qualities such men may possess, nature has not formed them for the command of armies, or the direction of great military operations.

LXXIV

To know the country thoroughly; to be able to conduct a *reconnaissance* with skill; to superintend the transmission of orders promptly; to lay down the most complicated movements intelligibly, but in a few words and with simplicity: these are the leading qualifications which should distinguish an officer selected for the head of the staff.

LXXV

A COMMANDANT of artillery should understand well the general principles of each branch of the service, since he is called upon to supply arms and ammunition to the different corps of which it is composed. This correspondence with the commanding officers of artillery at the advanced posts should put him in possession of all the movements of the army, and the disposition and management of the great park of artillery should depend upon this information.

LXXVI

To reconnoitre accurately defiles and fords of every description. To provide guides that may be depended upon. To interrogate the curé and postmaster. To establish rapidly a good understanding with the inhabitants. To send out spies. To intercept public and private letters. To translate and analyse their contents. In a word, to be able to answer every question of the general-in-chief when he arrives at the head of the army; these are the qualities which distinguish a good general of advanced posts.

LXXVII

GENERALS-in-chief must be guided by their own experience or their genius. Tactics, evolutions, the duties and knowledge of an engineer or an artillery officer may be learned in treatises, but the science of strategy is only to be acquired by experience, and by studying the campaigns of all the great captains.

Gustavus Adolphus, Turenne, and Frederick, as well as Alexander, Hannibal, and Cæsar, have all acted upon the same principles. These have been: to keep their forces united; to leave no weak part unguarded; to seize with rapidity on important points.

Such are the principles which lead to victory, and which, by inspiring terror at the reputation of your arms, will at once maintain fidelity and secure subjection.

LXXVIII

Peruse again and again the campaigns of Alexander, Hannibal, Cæsar, Gustavus Adolphus, Turenne, Eugene and Frederick. Model yourself upon them. This is the only means of becoming a great captain, and of acquiring the secret of the art of war. Your own genius will be enlightened and improved by this study, and you will learn to reject all maxims foreign to the principles of these great commanders.

ORIGINAL ANNOTATIONS
supplemented and updated
by David G. Chandler

I

The frontiers of states are either large rivers, or chains of mountains, or deserts. Of all these obstacles to the march of an army, the most difficult to overcome is the desert; mountains come next, and large rivers occupy the third place.

NAPOLEON in his military career appears to have been called upon to surmount every species of difficulty peculiar to aggressive warfare.

In Egypt he traversed deserts and vanquished and destroyed the Mamelukes, so celebrated for their address and courage. His genius knew how to accommodate itself to all the dangers of this distant enterprise in a country ill adapted to supply the wants of his troops.

In the conquest of Italy he twice crossed the Alps by the most difficult passes, and at a season which rendered this undertaking still more formidable. In three months he passed the Pyrenees, beat and dispersed four Spanish armies. In short, from the Rhine to the Borysthenes no natural obstacle could be found to arrest the rapid march of his victorious army.

CHANDLER: Neither Napoleon nor his editor take into account perhaps the must daunting state frontier of all (particularly for a continental power): a hostile sea coast. The problems involved in launching a major amphibious operation against Great Britain proved just as daunting and insuperable to France in 1797–8 and 1804–5 as they did to Hitler in 1940. In both cases the Royal Navy played a major role in ensuring England's national inviolability. In 1804–5 it was "the distant line of storm-toss'd wooden ships, upon which the Grand Army never gazed, that stood between Napoleon and the dominion of the world" (Mahan). In 1940 it was the Home Fleet, based upon Scapa Flow, and the East Coast and Channel flotillas of lighter naval shipping, that helped hold Hitler at bay. Although some of his generals stated that crossing the Channel would be little more than stepping over a puddle, the admirals of the *Kreigsmarine* knew better. Even more important, the air element (not present in the 1800s except in the imagination of certain artists who painted balloon-ships and one-man balloonists, together with a Channel Tunnel, as possible means of breaching England's south-coast defences) played the most critical role of all in 1940. It was the *Luftwaffe*'s failure to win the vital Battle of Britain against the RAF in the summer and autumn of 1940 – and even more important its chiefs' decision to switch their bombing effort against London just at the moment when the airfields and radar stations near the coasts of Kent and Sussex were about to crack – that arguably cost Hitler the Second World War. "Never in the field of human conflict", claimed Winston Churchill of his few hundred fighter pilots, "has so much been owed by so many to so few."

Land frontiers continue to impose important problems to attacking armies. Although the vaunted Maginot Line failed to check the German advance in 1940 (the divisions simply turned its flank through the Ardennes), the crossing of the Rhine in 1945 proved no easy task for the Allies despite the unforeseen gift of a captured intact bridge at Remagen in early March. In modern guerrilla wars, frontiers often provide the revolutionaries with sanctuaries in neutral territory (as was the case of the southern Thai borders with Malaya in the Emergencies of 1948–60 and 1968–75), or with vital lines of communication and supply (as in the case of the Ho Chi Minh Trail through neutral Laos and Cambodia during the Vietnam War).

II

In forming the plan of a campaign, it is requisite to foresee everything the enemy may do, and to be prepared with the necessary means to counteract it. Plans of campaign may be modified ad infinitum *according to circumstances, the genius of the general, the character of the troops, and the features of the country.*

SOMETIMES we see a hazardous campaign succeed, the plan of which is directly at variance with the principles of the art of war. But this success depends generally on the caprice of fortune, or upon faults committed by the enemy – two things upon which a general must never count. Sometimes the plan of a campaign runs the risk of failing at the outset, if opposed by an adversary who acts at first on the defensive, and then suddenly seizing the initiative, surprises by the skilfulness of his manœuvres.

Such was the fate of the plan laid down by the Aulic Council for the campaign of 1796, under the command of Marshal Würmser. From his great numerical superiority, the marshal had calculated on the entire destruction of the French army by cutting off its retreat. He founded his operations on the defensive attitude of his adversary, who was posted on the line of the Adige, and had to cover the siege of Mantua, as well as central and lower Italy.

Würmser, supposing the French army fixed in the neighbourhood of Mantua, divided his force into three corps which marched separately, intending to unite at that place. Napoleon, having penetrated the design of the Austrian general, felt all the advantage to be derived from striking the first blow against an army divided into three corps without any relative communications. He hastened, therefore, to raise the siege of Mantua, assembled the whole of his forces, and by this means became superior to the Imperialists, whose divisions he attacked and beat in detail. Thus Würmser, who fancied he had only to march to certain victory, saw himself compelled after a ten days' campaign to retire with the remains of his army into the Tyrol, after a loss of twenty-five thousand men in killed and wounded, fifteen thousand prisoners, nine stand of colours, and seventy pieces of cannon. This proves that nothing is so difficult as to prescribe beforehand to a general the line of conduct he shall pursue during the course of a campaign. Success must often depend on circumstances that cannot be foreseen; and it should be remembered, likewise, that nothing cramps so much the efforts of genius as compelling the head of an army to be governed by any will but his own.

CHANDLER: This remains as true as ever of any modern war. "When making plans," Churchill once remarked, "it is as well to take into account those of the enemy." This maxim connotes the vital importance of military intelligence gathering and evaluation. Today this is a huge and complex matter involving space satellites, electronic sensors and detectors of all kinds, as well as old-fashioned spying by human observation. The huge mass of information available makes its sorting and evaluation very difficult – as the Vietnam War demonstrated, when the Americans received several tons of paper intelligence reports each day.

A classic case of failing to observe this maxim was the Falklands War of 1982. The Argentinians indubitably achieved strategic surprise by seizing the islands when they did – but then they fatally miscalculated the determination of the British response. Despite a considerable period for counter-invasion preparations, they failed to make the islands impregnable against the British Expeditionary Task Force. The effects of bombs and French-supplied Exocet airborne missiles proved unexpectedly grave for the vulnerable British frigates and supply ships (one thinks of HMS *Sheffield* and HMS *Ardent*, *Sir Galahad* and *Atlantic Conveyor*), but it was the superiority in the air – aircraft to aircraft – enjoyed by the British Harriers and the tough fighting skills of regular paratroopers and Royal Marines pitted mainly against raw Argentinian conscripts, and above all the firm and unwavering resolve of the Thatcher government at the political level that General Galtieri and his advisers failed to take into account or prepare adequately for.

Obviously flexibility of planning remains absolutely vital, taking into account the same factors that Napoleon mentioned. Thus the accurate and flexible Israeli assessments of their various Arab opponents' capabilities underlay their successes in the Six Day War and the Yom Kippur War – although on the latter occasion the Egyptians gained initial surprise by the timing and form of their attack over the Suez Canal against the under-manned Bar-Lev Line. But the Israelis then reacted strongly and highly effectively. In overall terms, therefore, Napoleon's maxim that what is needed is "a plan of many branches" remains as true today as in the 1800s. The complexity and flexibility at the security and operational levels of Operation "Overlord" against German-dominated Europe in 1944 is a good case in point.

III

An army which undertakes the conquest of a country has its two wings either resting upon neutral territories or upon great natural obstacles, such as rivers or chains of mountains. It happens in some cases that only one wing is so supported, and in others that both are exposed. In the first instance . . . But in all these cases it is necessary every five or six days to have a strong post, or an entrenched position, upon the line of march. . . .

THESE general principles in the art of war were entirely unknown or lost sight of in the Middle Ages. The Crusaders, in their incursions into Palestine, appear to have had no object but to fight and to conquer, so little pains did they take to profit by their victories. Hence innumerable armies perished in Syria, without any other advantage than that derived from the momentary success obtained by superior numbers.

It was by neglect of these principles also, that Charles the Twelfth, abandoning his line of operation and all communication with Sweden, threw himself into the Ukraine, and lost the greater part of his army by the fatigues of a winter campaign in a barren country destitute of resources.

Defeated at Pultawa, he was reduced to seek refuge in Turkey, after crossing the Nieper with the remains of his army, diminished to little more than one thousand men.

Gustavus Adolphus was the first who brought back the art of war to its true principles. His operations in Germany were bold, rapid, and well executed. He made success at all times conducive to future security, and established his line of operation so as to prevent the possibility of any interruption in his communications with Sweden. His campaigns form a new era in the art of war.

CHANDLER: The significance of this maxim has lost nothing over the past 86 years, but important new applications have made their appearance. The first paragraph recalls the use of neutral Laos and Cambodia by the Vietcong and North Vietnamese for their famous line of communication, the Ho Chi Minh Trail. The Pentagon recognised this primitive route's crucial importance and launched heavy bombing raids (Operation "Rolling Thunder") against it, and in the spring of 1970 President Nixon sanctioned the incursion by ARVN forces (with American advisers in support) into Cambodia. This decision – designed to earn a six-month respite for the withdrawal of US troops and a relatively quiescent take-over period for the South Vietnamese government forces – backfired psychologically and politically, triggering off a wave of American university campus revolts and in effect breaking the determination of the American government and people to seek a successful outcome to the limited war their country was fighting in South-East Asia.

Similarly, the need under the third instance cited in the maxim for "all-round defence" at the operational level when both flanks of a force are exposed is well illustrated by events in Burma, the Western Desert, and the USSR during the Second World War. The "Battle of the Admin. Box" (1943) and the herculean struggles around Kohima and Imphal (1944) are examples of General Slim's employment of the method in Burma. The use of self-contained all-round defence positions, protected by minefields, as in the "Knightsbridge Box" and other positions of the ill-fated Gazala Line by General Auchinleck in Libya (1942) is another significant example. And the fate of von Paulus's entrapped German Sixth Army at Stalingrad (1942–3) and the skilful use of small defended localities by the great Manstein to delay the Soviet Red Army's exploitation of its victory at Kursk (1943–4), is a third instance. Of course two vital developments since 1901 have been the widespread use of wire and minefields for protecting defensive positions and the coming of air power and, above all in this case, air supply – the factor that let down von Paulus and permitted Slim to triumph.

The point about placing strong "Forward Supply Depots" behind advancing armies at regular intervals remains equally important today – for fuel, munitions and supplies. Wellington was careful in the Peninsular War to set up depots every 50 miles behind his front, running back to his main base at Lisbon. In 1940–3 the same rule applied in the Western Desert, as illustrated by General O'Connor's advance in pursuit of the Italian Tenth Army in late 1940, his communications running back to Mersa Matruh and thence to Alexandria and Cairo. The ox cart, pack mule and lorry have been replaced – in whole or in part – by air supply (fly-in, air-drop or helicopter lift) today, although motor convoys remain much used for the support of large operations. Both supply depots and convoys remain major targets for interdictive air or land attack.

It should be noted that a "line of operations" extends ahead of an army in the field; its "line of communications" runs to its rear. The ideal positioning of the latter to the FEBA (Forward Edge of the Battle Area) is at right angles to the front so as to achieve maximum security.

IV

When the conquest of a country is undertaken by two or three armies, which have each their separate line of operation until they arrive at a point fixed upon for their concentration, it should be laid down as a principle, that the junction should never take place near the enemy. . . .

IN the campaign of 1757, Frederick, marching to the conquest of Bohemia with two armies, which had each their separate line of operation, succeeded, notwithstanding, in uniting them in sight of the Duke of Lorraine, who covered Prague with the Imperial army. The success of this march, however, depended entirely on the inaction of the duke, who at the head of seventy thousand men did nothing to prevent the junction of the two Prussian armies.

CHANDLER: "March dispersed, fight concentrated" is one of the best known Napoleonic maxims. The need to concentrate *off* the battlefield so as to avoid the risk of piecemeal destruction is also an important tenet. However, the master broke the rule himself at Marengo (1800), where he allowed his army to be attacked by General Melas while two large French detachments (Desaix's and Lapoype's) were absent at the outset. The fate of Blucher's Prussians on 16 June 1815, forced to fight at Ligny while still in the process of concentrating forward, is a good example, and military history holds many another case. Sometimes success may be gained by the sheer surprise of undertaking such a manoeuvre in the foe's proximity – thus the Duke of Marlborough deliberately fed his army piecemeal over the River Scheldt into battle at Oudenarde (1708), but admitted next day that only the desperate need for a decisive battle persuaded him to risk it. "This only made me venture the battle yesterday; otherwise I did give them too much advantage." This illustrates the point that maxims – like principles of war – are NOT intended to be hard-and-fast rules determining conduct; rather they are guides to thought processes and planning. Circumstances alter cases.

Possibly the greatest example of Napoleon's putting his maxim into effect was the advance from the Rhine to the Danube in the late summer of 1805. The Grand Army of 210,000 men moved through Germany in seven separated but steadily converging columns before concentrating around Dönauwörth for the Danube crossing prior to sweeping on to envelop "the unfortunate General Mack" at Ulm. Moltke the Elder's advance upon the Austrians before Königgrätz in 1866 is another example. Today, however, the possible use of nuclear tactical weapons on the NATO fronts makes all concentrations of force except in close proximity to the enemy potentially hazardous unless these can be concealed from the foe. Thus concentrations of force when such weaponry might be employed must be for very brief periods only. Napoleon – although he once stated that the use of "thunderbolts" where possible were to be preferred to that of cannon – did not in this case think through the implications of full concentration before battle. It is relevant to note that Napoleon meant "assembled" rather than "concentrated" in this case. He made great play of bringing up uncommitted corps to outflank the enemy. Everything depended on gaining and keeping the initiative. Without it the penalties he describes may well occur.

V

All wars should be governed by certain principles, for every war should have a definite object, and be conducted according to the rules of art. . . .

IT was a saying of Marshal Villars, that when war is decided on, it is necessary to have exact information of the number of the troops the enemy can bring into the field, since it is impossible to lay down any solid plan of offensive or defensive operation without an accurate knowledge of what you have to expect and fear. When the first shot is fired, no one can calculate what will be the issue of the war. It is therefore of vast importance to reflect maturely before we begin it. When once, however, this is decided, the marshal observes that the boldest and most extended plans are generally the wisest and the most successful. When we are determined upon war, he adds, we should carry it on vigorously and without trifling.

CHANDLER: The need for a clear aim in any way is self-evident (and arguably was lacking when the USA intervened in strength in Vietnam). This maxim has been taken to imply slavish adherence to the so-called "Principles of War" – surprise, concentration of force, flexibility, security, morale and the rest (the lists have varied and still vary between countries and periods), and in particular to that of "Economy of Force". In fact such an attitude is almost invariably fatal; as mentioned above, "principles" are supposed to be "guides to possible conduct", never hard and fast rules. There is, however, some general acceptability about the need to tailor "means" to "ends" in any war. To engage too many forces for a particular purpose is clearly wasteful of resources. The American "liberation" of Grenada in the West Indies (1983) is arguably a case of "overkill" or using a "sledgehammer to crack a nut". Conversely, to go into a war with insufficient forces is obviously fool-hardy in the extreme, although, as the Israelis found in 1948 and 1967, circumstances may compel this – and skill can offset the dangers. The British commitment of naval, air and land forces to the reconquest of the Falklands in 1982 was well considered, although the naval aspect was "a near-run thing" and a calculated risk, as the loss of five warships, and above all the critically important logistics vessel *Atlantic Conveyor* to air attack well demonstrated. From first to last, however, Mrs Thatcher's aim was crystal clear – and is being maintained to this day. Clearly all recourse to military force at whatever level or scale needs the most careful consideration, political as well as military. Napoleon was generally extremely astute at this – one of the functions of Grand Strategy – although twice he made devastating miscalculations: in 1807–8 by bringing on the Peninsular War in Portugal and Spain (and then persisting in it to the bitter end); and in 1812 by undertaking the invasion of Russia. General Galtieri clearly miscalculated the likely British response in the South Atlantic in 1982, although his aim was clear enough; just as forty-or-so years earlier another dictator, Adolf Hitler, fatally underestimated Winston Churchill's and the British people's resolve to defy him to the bitter end. Similarly the French and later the Americans may be said to have consistently underestimated the determination of Ho Chi Minh and General Giap in Indo-China/Vietnam. On the other hand, the Israelis have so far proved highly adept at adjusting means to ends in the long series of Middle

East Wars. Above all, this ability depends on securing sufficient accurate intelligence – and interpreting it correctly.

VI

At the commencement of a campaign, to advance *or* not to advance *is a matter for grave consideration, but when once the offensive has been assumed, it must be sustained to the last extremity.* . . .

MARSHAL Saxe remarks that no retreats are so favourable as those which are made before a languid and unenterprising enemy, for when he pursues with vigour, the retreat soon degenerates into a rout. Upon this principle it is a great error, says the Marshal, to adhere to the proverb which recommends us to build a bridge of gold for a retreating enemy. No – follow him up with spirit, and he is destroyed.

CHANDLER: This maxim illustrates Napoleon's conviction that a "rapid and skilful offensive" must grow out of a "careful, and circumspect, defensive". In other words an attack must be based on careful preparations and not be rashly entered upon without taking necessary preliminary precautions. This, as in so many instances, is often easier said than done. Once entered upon, however, an attack must be boldly pressed – or, in Guderian's words concerning North-West Europe (1940), an army must "head for the terminus". Indecision and hesitation will play into the enemy's hands, help him to recover and even permit him to turn the tables.

The advantages in morale terms of being on the offensive rather than the defensive are clear. In 1940 it was their "Maginot Line complex" and lack of all initiative during the preceding winter that doomed the French to cataclysmic defeat. However, there are indications that NATO's Central Front philosophy of waiting to be attacked, and when the attack comes of conceding ground, is potentially very dangerous. Although in some ways it is easier to fight on the defensive, the morale of conscript soldiers in particular seems to falter as they find themselves surrendering their own countryside in order to gain a little time. On the other hand it can be argued – as in the case of the Southern States in the American Civil War or of the Israeli forces in their struggles since 1948 – that "the defence of the Motherland" has a most powerful emotive effect. As always, therefore, there are two sides to the argument. Thus in a campaign such as the Western Desert (1940–2), where the main British and German contenders were not fighting in their native countries, the effects of a skilful retreat can be the reverse of those described here by Napoleon. The retiring army is falling back upon its line of communications, absorbing depots, reinforcements and the like in the process, while its triumphant pursuer is getting increasingly over-extended and more and more distant from his bases. Eventually, as happened to Napoleon in Russia (1812), the laws of "strategic consumption" slow down the advance and eventually halt it, provided that the "defeated" side maintains its willpower and its faith in ultimate success. But Napoleon is right in asserting that a retiring force may lose much material. After great tank engagements – such as "Knights-bridge" (1942) or Kursk (1943) – the victors can recover and repair their damaged vehicles whereas the losers cannot. A recent example of this was the Libyan losses in armour and military stores sustained after their defeat at Ouadi Doum in northern Chad (1987).

VII

An army should be ready every day, every night, and at all times of the day and night, to oppose all the resistance of which it is capable. . . .

THE following maxims, taken from the memoirs of Montecuccoli, appear to me well suited to this place, and calculated to form a useful commentary on the general principles laid down in the preceding observations: –

1

When war has been once decided on, the moment is past for doubts and scruples. On the contrary, we are bound to hope that all the evil which may ensue, will not; that Providence, or our own wisdom, may avert it, or that the want of talent on the part of the enemy may prevent him from benefiting by it. The first security for success is to confer the command on one individual. When the authority is divided, opinions are divided likewise, and the operations are deprived of that *ensemble* which is the first essential to victory. Besides, when an enterprise is common to many, and not confined to a single person, it is conducted without vigour, and less interest is attached to the result.

After having strictly conformed to all the rules of war, and satisfied ourselves that nothing has been omitted to ensure eventual success, we must then leave the issue in the hands of Providence, and repose ourselves tranquilly in the decision of a higher power.

Let what will arrive, it is the part of a general-in-chief to remain firm and constant in his purposes: he must be equally superior to elation in prosperity and depression in adversity, for in war good and bad fortune succeed each other by turns, and form the ebb and flow of military operations.

2

When your own army is strong and inured to service, and that of the enemy weak and consisting of new levies, or of troops enervated by long inaction, then you should exert every means to bring him to battle.

If, on the other hand, the adversary has the advantage in troops, a decisive combat is to be avoided, and you must be content to impede his progress, by encamping advantageously and fortifying favourable passes. When armies are nearly equal, it is desirable *not* to avoid a battle, but only to fight one to advantage. For this purpose care should be taken to encamp always in front of the enemy, to move when he moves, and occupy the heights and advantageous grounds that lie upon his line of march; to seize upon all the buildings and roads adjoining to his camp, and post yourself advantageously in the places by which he must pass. It is always something gained to make *him* lose time, to thwart his designs, or to retard their progress and execution. If, however, an army is altogether inferior to that of the enemy, and there is no possibility of manœuvring against him with success, then the campaign must be abandoned, and the troops must retire into the fortresses.

3

The first object of a general-in-chief in the moment of battle should be to secure the flanks of his army. It is true that natural positions may be found to effect this object, but these positions being fixed and immovable in themselves, they are only advantageous to a general who is prepared to wait the shock of the enemy, and not to one who marches to the attack.

A general can therefore rely only on the just disposition of his troops to enable him to repel any attempt the adversary may make upon the front, or flanks, or rear of his army.

If one flank of an army rests upon a river, or an impassable ravine, the whole of the cavalry should be posted with the other wing, in order to envelop the enemy more easily by its superiority in numbers.

If the enemy has his flanks supported by woods, light cavalry or infantry should be despatched to attack him in flank, or in rear, during the heat of the battle. If practicable, also, an attack should be made upon the baggage, to add to his confusion.

If you desire to beat the enemy's left with your right wing, or his right with your left wing, the wing with which you attack should be reinforced by the *élite* of your army. At the same moment the other wing should be refused, and the attacking wing brought rapidly forward, so as to overwhelm the enemy. If the nature of the ground admits, he should be approached by stealth, and attacked before he is on his guard. If any signs of fear are discoverable in the enemy, and which are always to be detected by confusion or disorder in his movements, he should be pursued immediately, without allowing him time to recover himself. It is now the cavalry should be brought into action, and manœuvre so as to cut off his artillery and baggage.

4

The order of march should always be subservient to the order of battle, which last should be arranged beforehand. The march of an army is always well regulated when it is governed by the distance to be accomplished, and by the time required for its performance. The front of the column of march should be diminished or increased according to the nature of the country, taking care that the artillery always proceeds by the main road.

When a river is to be passed, the artillery should be placed in battery upon the bank opposite the point of crossing.

It is a great advantage when a river forms a sweep or angle, and when a ford is to be found near the place where you wish to effect a passage. As the construction of the bridge proceeds, infantry should be brought to cover the workmen by keeping up a fire on the opposite bank; but the moment it is finished, a corps of infantry and cavalry and some field-pieces should be pushed across. The infantry should entrench itself immediately at the head of the bridge, and it is prudent, moreover, to fortify on the same side of the river, in order to protect the bridge, in case the enemy should venture an offensive movement.

The advanced guard of an army should be always provided with trusty guides and with a corps of pioneers – the first to point out the best road, the second to render bad ones more practicable.

If the army marches in detachments, the commander of each detachment should be furnished with the name of the place in writing where the whole are to be reassembled. The place should be sufficiently removed from the enemy to prevent him from occupying it before the junction of all the detachments. To this end it is of importance to keep the name a secret.

From the moment an army approaches the enemy it should march in the order in which it is intended to fight. If anything is to be apprehended, precautions are necessary in proportion to the degree of the danger. When a defile is to be passed, the troops should be halted beyond the extremity, until the whole army has quitted the defile.

In order to conceal the movements of an army, it is necessary to march by night, through woods and valleys, by the most retired roads, and out of reach of all inhabited places. No fires should be allowed, and to favour the design still more, the troops should move by verbal order. When the object of the march is to carry a post, or to succour a place that is besieged, the advanced guard should march within musket-shot of the main body, because then you are prepared for an immediate attack, and ready to overthrow all before you.

When a march is made to force a pass guarded by the enemy, it is desirable to make a feint upon one point, while by a rapid movement you bring your real attack to bear upon another.

Sometimes success is obtained by pretending to fall back upon the original line of march, and by a sudden countermarch seizing upon the pass before the enemy is able to reoccupy it. Some generals have gained their point by manœuvring so as to deceive the enemy, while a detachment under the cover of high grounds has surprised the passage by a stolen march. The enemy being engaged in watching the movements of the main body, the detachment has an opportunity of entrenching itself in its new position.

5

An army regulates its mode of encampment according to the greater or less degree of precaution which circumstances require. In a friend's country the troops are divided to afford better accommodation and supplies. But with the enemy in front, an army encamps always in order of battle. With this view, it is of the last importance to cover one part of the camp as far as practicable by natural defences, such as a river, a chain of rocks, or a ravine. Care should be taken also that the camp is not commanded, and that there is no obstacle to a free communication between the different corps, and which can prevent the troops from mutually succouring each other.

When an army occupies a fixed camp, it is necessary to be well supplied with provisions and ammunition, or at least that these should be within certain reach and easily obtained. To ensure this, the line of communication must be well established, and care taken not to leave an enemy's fortress in your rear.

When an army is established in winter quarters, its safety is best secured either by fortifying a camp (for which purpose a spot should be selected near a large commercial town, or a river affording facility of transport), or by distributing it in close cantonments, so that the troops should be near together, and capable of affording each other mutual support.

The winter quarters of an army should be protected likewise by constructing small covered works on all the lines of approach to the cantonments, and by posting advanced guards of cavalry to observe the motions of the enemy.

6

A battle is to be sought when there is reason to hope for victory, or when an army runs the risk of being ruined without fighting; also when a besieged place is to be relieved, or when you desire to prevent a reinforcement from reaching the enemy. Battles are useful likewise when we wish to profit by a favourable opportunity which offers to secure a certain advantage, such as seizing upon an undefended point or pass, attacking the enemy when he has committed a fault, or when some misunderstanding among his generals favours the undertaking.

If an enemy declines an engagement, he may be compelled to it, either by besieging a place of importance or by falling upon him unawares, and when he cannot easily effect his retreat, or (after pretending to retire) by making a rapid counter-march, attacking him vigorously, and forcing him to action.

The different circumstances under which a battle should be avoided or declined are, when there is greater danger to be apprehended from defeat than advantage to be derived from victory; when you are very inferior to your enemy in numbers, and are expecting reinforcements; above all, when the enemy is advantageously posted, or when he is contributing to his own ruin by some inherent defect in his position or by the errors and divisions of his generals.

To gain a battle, each arm must be advantageously posted, and have the means of engaging to its front and flank. The wings must be protected by natural obstacles where these present themselves, or by having recourse, when necessary, to the aid of art.

The troops must be able to assist each other without confusion, and care must be taken that the broken corps do not fall back upon and overthrow the rest. Above all, the intervals between the different bodies must be sufficiently small to prevent the enemy from penetrating, for in that case you would be obliged to employ your reserves, and run the risk of being entirely overwhelmed.

Sometimes victory is obtained by creating a diversion in the middle of a battle, or even by depriving the soldier of all hope of retreat, and placing him in a situation where he is reduced to the necessity either to conquer or die. At the commencement of a battle, if the ground is level, you should advance to meet the enemy, in order to inspire the soldier with courage; but if you are well posted, and your artillery advantageously placed, then wait for him with determination, remembering always to fight resolutely, to succour opportunely those who require it, and never to commit your reserves except in the last extremity, and even then to preserve some support behind which the broken corps may rally.

When it is necessary to attack with your whole force, the battle should commence towards evening, because then, whatever be the issue, night will arrive to separate the combatants before your troops are exhausted. By this means an opportunity is afforded of effecting an orderly retreat if the result of the battle requires it.

During an action the general-in-chief should occupy some spot from whence he can, as far as possible, over-look his whole army. He should be informed immediately of everything that passes in the different divisions. He should be ready, in order to render success more complete, to operate with fresh troops upon those points where the enemy is giving way, and also to reinforce his own corps wherever they are inclined to yield. When the enemy is beaten, he must pursue him instantly, without giving him a moment to rally; on the other hand, if he is himself defeated or despairs of victory, he must retire betimes in the best possible order.

7

It shows great talent in a general to bring troops who are prepared for action into collision with those who are not; for example, fresh troops against those which are exhausted, brave and disciplined men against recruits. He must likewise be ready always to fall with his army upon a weak or detached corps, to follow the track of the enemy, and charge him among defiles before he can face about and get into position.

8

A position is good when the different arms are so placed as to be engaged with advantage, and without any remaining unemployed. If you are superior in cavalry, positions are to be sought in plains and open ground. If in infantry, in an enclosed and covered country. If inferior in numbers, in confined and narrow places; if superior, in a spacious and extensive field. With a very superior army, a difficult pass must be selected to occupy and fortify.

9

In order to obtain every possible advantage from a diversion, we should ascertain first that the country in which it is to be created is easily penetrated. A diversion should be made vigorously, and on those points where it is calculated to do the greatest mischief to the enemy.

10

To make war with success the following principles should never be departed from: –

To be superior to your enemy in numbers as well as in *moral*; to fight battles in order to spread terror in the country; to divide your army into as many corps as may be effected without risk, in order to undertake several objects at the same time; to treat *well* those who yield, to *ill*-treat those who resist; to secure your rear, and occupy and strengthen yourself at the outset in some post which shall serve as a central base point for the support of your future movements; to make yourself master of the great rivers and principal passes, and to establish your line of communication by getting possession of the fortresses by laying siege to them, and of the open country by giving battle; for it is vain to expect that conquests are to be achieved without combats, although when the victory is won they will be best maintained by uniting mildness with valour.

CHANDLER: The need for constant vigilance on active service is even more necessary today than it was in Napoleonic times. Modern radar, satellite surveillance and night-vision equipment – ranging from "artificial moonlight" to infra-red and heat-source sights – imply that action can be continuous with scant pauses unless these are dictated by resupply problems or sheer exhaustion. This applies as much to fighting in the air and at sea as it does to land operations. Security is a vital consideration at all times; hence the need for aggressive patrolling, continuous reconnaissance, and ceaseless updating of plans and intentions. Electronic surveillance makes secrecy hard to achieve, although countermeasures such as jamming devices have also been brought to a very developed state. All due precautions need to be taken at all times to avoid being surprised – that is the nub of Napoleon's thinking here.

In all other respects, the extensive notes by the original editor are still very relevant to the modern commander; however, Montecuccoli's advice in paragraph 10 above to ill-treat those who resist needs qualification in the modern age.

VIII

A general-in-chief should ask himself several times in the day, What if the enemy were to appear now in my front, or on my right, or my left?

In the campaign of 1758 the position of the Prussian army at Hohen Kirk, being commanded by the batteries of the enemy who occupied all the heights, was eminently defective. Notwithstanding, Frederick, who saw his rear menaced by the corps of Laudon, remained six days in his camp without seeking to correct his position. It would seem, indeed, that he was ignorant of his real danger, for Marshal Daun, having manœuvred during the night in order to attack at daybreak, surprised the Prussians in their lines before they were able to defend themselves, and by this means surrounded them completely.

Frederick succeeded, however, in effecting his retreat with regularity, but not without the loss of ten thousand men, many general officers, and almost all his artillery.

If Marshal Daun had followed up his victory with greater boldness, the King of Prussia would never have been able to rally his army. On this occasion Frederick's good fortune balanced his imprudence.

Marshal Saxe remarks that there is more talent than is dreamt of in bad dispositions if we possess the art of converting them into good ones, when the favourable moment arrives. Nothing astonishes the enemy so much as this manœuvre. He has counted upon *something*. All his arrangements have been founded upon it accordingly; and at the moment of attack it escapes him! "I must repeat," says the marshal, "there is nothing that so completely disconcerts an enemy as this, or engages him to commit so many errors; for it follows that, if he does *not* change his dispositions, he is beaten; and if he *does* change them, in presence of his adversary, he is equally undone."

It seems to me, however, that a general who should rest the success of a battle upon such a principle, would be more likely to lose than to gain by it; for, if he had to deal with a skilful adversary and an alert tactician, the latter would find time to take advantage of the previous bad dispositions before he would be able to remedy them.

CHANDLER: Obviously, a careful commander-in-chief (and, indeed, a commander at any level) should remain alert to all possibilities and strive to ensure that his position is sufficiently well balanced to permit a quick and effective reaction to the unexpected. As Clausewitz strongly claims, "war is the province of uncertainty and friction". The suitability of a position for all-round defence at short or no notice is clearly a desirable characteristic, particularly today with the possibility of sudden attack by heli-borne forces. It also applies strongly to the requirements for security in guerrilla-type operations. A late 20th century requirement that did not apply in Napoleonic or Boer War times is that of camouflage against air attack. Track-discipline, proper dispersal of vehicles and disposal of spoil from digging trenches have been vital since at least as far back as 1939.

Another side of the same coin is the need for a commander, even when not in a dangerous situation, to spend much time thinking: "How would I exploit the possibilities, and avoid the disadvantages, of this piece of terrain around me?" Wellington called it "getting to know what is on the other side of the hill and learning what we don't know by the study of what to do". It is often remembered that he had looked over the battlefield area of Waterloo in 1814 and had stored it away in his mind as a possible position; it is less often recalled that in fact he intended early on 17 June 1815 to fall back from Quatre Bras not to the ridge of Mont St Jean but to the one immediately to its south – the position taken up by the French army that evening. The decision to take post on the more northerly ridge, with its back to the Bois de Soignies, was taken by his Quartermaster-General, de Launcey, when he rode back ahead of the retiring Allied army to take a hard look at the position.

IX

The strength of an army, like the power in mechanics, is estimated by multiplying the mass by the rapidity. . . .

RAPIDITY, says Montecuccoli, is of importance in concealing the movements of an army, because it leaves no time to divulge the intention of its chief. It is, therefore, an advantage to attack the enemy unexpectedly, to take him off his guard, to surprise him, and let him feel the thunder before he sees the flash. But if too great celerity exhausts your troops, while, on the other hand, delay deprives you of the favourable moment, you must weigh the advantage against the disadvantage and choose between. Marshal Villars observes, that in war everything depends upon being able to deceive the enemy, and, having once gained this point, in never allowing him time to recover himself. Villars has united practice to precept. His bold and rapid marches were almost always crowned with success. It was the opinion of Frederick that all wars should be short and rapid; because a long war insensibly relaxes discipline, depopulates the state, and exhausts its resources.

CHANDLER: Napoleon was the master of the sudden dash designed to disconcert the enemy and morally dominate him. A century earlier Marlborough had been adept at the same method. "If he is there then the Devil must have carried him," was Vendôme's reputed reaction to the Duke's sudden advance towards Oudenarde in July 1708; "Such marching is impossible!" The celebrated ability of Erwin Rommel to "steal a march" on his opponents was as manifest when he was in command of the "Ghost" Panzer Division during the breakthrough over the Meuse at Sedan in May 1940 as when he commanded the *Afrika Korps* in the Western Desert in 1941 and 1942.

The exhilaration of a rapid advance as a multiplier of battle force and troop morale has never been better epitomised than in the Red Army's dramatic offensive in August 1945 against the Japanese Imperial Army in

Mongolia. In three weeks fighting some 500 miles were covered and one million Japanese defeated. That is the type of *blitzkrieg* attack that modern Soviet officers are being encouraged to seek to emulate in any future war. Clearly morale is likely to rise in any force making ground (as has already been mentioned at VI above). The argument is sometimes heard concerning NATO's Central Front that consideration should be given to the idea of getting in the first blow when politically the outbreak of hostilities has become certain, and attacking to the East of the inner German border before the Warsaw Pact forces are fully ready. Such an aggressive, preemptive move would have obvious morale advantages – if successful – and the West Germans would not entirely object to having a future war fought over the frontier in East Germany and Czechoslovakia rather than on their own soil. But NATO's role as an "Alliance for Peace" is a vital political consideration and carries the explicit assumption that armed action would only be triggered by an enemy attack, thus effectively sacrificing the initiative at the outset. But, as the Israelis demonstrated in 1967, a pre-emptive strike (in that case against the unwary Egyptian air force) followed by an all-out attack with air-supported armour into Sinai (leading to a redeployment north to meet the Syrians on the Golan Heights) had much to recommend it as a way of compensating for inferior strength. President Nasser had "sown the wind", as Israelis see it, by requiring the United Nations observer force to withdraw, and Egypt certainly "reaped the whirlwind" in the Six Day War that resulted. In this case a rapid advance clearly both increased the chances of victory and boosted Israeli morale to new heights.

X

When an army is inferior in number, inferior in cavalry and in artillery, it is essential to avoid a general action. . . .

THE campaign of 1814 in France was skilfully executed upon these principles. Napoleon, with an army inferior in numbers, an army discouraged by the disastrous retreats of Moscow and of Leipzig, and still more by the presence of the enemy in the French territory, contrived, notwithstanding, to supply his vast inequality of force by the rapidity and combination of his movements. By the success obtained at Champaubert, Montmirail, Montereau, and Rheims, he had already begun to restore the *moral* of the French army. The numerous recruits of which it was composed had already acquired that steadiness, of which the old regiments afforded them an example, when the capture of Paris and the astonishing revolution it produced compelled Napoleon to lay down his arms.

But this consequence resulted rather from the force of circumstances than from any absolute necessity, for Napoleon by carrying his army to the other side of the Loire might easily have formed a junction with the armies of the Alps and Pyrenees, and have reappeared on the field of battle at the head of a hundred thousand men. Such a force would have amply sufficed to re-establish the chances of war in his favour, more especially as the armies of the allied sovereigns were obliged to manœuvre upon the French territory with all the strong places of Italy and France in their rear.

CHANDLER: This maxim is again uncomfortably pertinent to the situation NATO finds itself facing on the Central Front or on the Flanks, where Soviet and Warsaw Pact superiority of strength in terms of conventional forces is notable. To give time for reinforcements to arrive from America (although substantial forces might well arrive by air during the "transition to war" phase) might require the Supreme Commander to sacrifice space for time in compliance with another famous Napoleonic dictum, "Ground I may recover, time never". However, as the Israelis have demonstrated time and again since 1948 on the Golan Heights, in Sinai and more recently in Lebanon, the use of surprise, fast movement and troops imbued with a high sense of morale can outweigh such disadvantages effectively.

This maxim is also highly applicable to the waging of Revolutionary Guerrilla Warfare. The avoidance of major actions during the second stage of Mao Tse Tung's celebrated three-part guerrilla war doctrine, and conducting operations based upon surprise, rapid hit-and-run tactics designed to increase the morale of the partisans and correspondingly to erode that of their opponents while impressing the general population with one's ability to attack a stronger enemy at will – these are all tenets of modern RGW. Equally, they figure large in the manuals on counter-insurgency (COIN).

XI

*To act upon lines far removed from each other, and without com-
munications, is to commit a fault which always gives birth to a
second. The detached column has only its orders for the first day.
Its operations on the following day depend upon what may have
happened to the main body. Thus the column on any sudden
emergency either loses time in waiting for orders or acts without
them and at hazard. . . .*

THE Austrian army, commanded by Field Marshal
Alvinzi, was divided into two corps, destined to act
independently, till they should accomplish their junction
before Mantua.

   The first of these corps, consisting of forty-five
thousand men, was under the orders of Alvinzi. It was to
debouch by Monte Baldo upon the positions occupied by
the French army on the Adige. The second corps, commanded by General Provéra, was destined to act upon
the lower Adige, and to raise the blockade of Mantua.
Napoleon, informed of the enemy's movements, but not
entirely comprehending his projects, confined himself to
concentrating his masses and giving orders to the troops
to hold themselves in readiness to manœuvre. In the
meantime fresh information satisfied the general-in-chief
of the French army that the corps which had debouched
by La Coronna over Monte Baldo was endeavouring to
form a junction with its cavalry and artillery, both
which, having crossed the Adige at Dolce, were directing
their march upon the plateau of Rivoli, by the great road
leading by Incanole.

Napoleon immediately foresaw that by having possession of the plateau he should be able to prevent this
junction, and obtain all the advantages of the initiative.
He accordingly put his troops in motion, and at two
o'clock in the morning occupied that important position.
Once master of the point fixed upon for the junction of
the Austrian columns, success followed all his dispositions. He repulsed every attack, made seven thousand
prisoners, and took several standards and twelve pieces
of cannon. At two o'clock in the afternoon the battle of
Rivoli was already gained, when Napoleon, learning that
General Provéra had passed the Adige at Anghiari, and
was directing his march upon Mantua, left to his
generals the charge of following up the retreat of Alvinzi,
and placed himself at the head of a division for the
purpose of defeating the designs of Provéra.

By a rapid march he again succeeded in the initiatory
movement, and in preventing the garrison of Mantua
from uniting its force with the relieving army. The corps
charged with the blockade, eager to distinguish itself
under the eyes of the conqueror of Rivoli, compelled the
garrison to retire into the place, while the division of
Victor, forgetful of the fatigues of a forced march,
attacked the relieving army in front. At this moment a
sortie from the lines of St. George took him in flank, and
the corps of Augereau which had followed the march of
the Austrian general attacked him in rear. Provéra, surrounded on all sides, capitulated. The result of these two
battles cost the Austrians three thousand men in killed
and wounded, twenty-two thousand prisoners, twenty-
four standards, and forty-six pieces of cannon.

CHANDLER: The perils of a force being caught dispersed in a number of ill-coordinated parts have already been discussed in Maxim IV above. Even in the early days of radio, the Russians were destroyed at Tannenberg in East Prussia (1914) by Hindenburg and Ludendorf by operating in two ill-coordinated wings. A century earlier, Wellington capitalised on Marshal Marmont's unwisely strung-out divisions striving to get ahead of him to block his line of retreat towards Portugal to fight the great battle of Salamanca. Clearly there are many times when detached corps need to be employed – especially in modern and future wars – but the requirements Napoleon lays down remain valid concerning a generally independent mandate for their commanders. One Second World War example would be the use of General Orde Wingate's "Chindits" in Burma (1943 and 1944). Another application would be the use of airborne attack – as at Arnhem in 1944 or at the Mitla Pass in 1967 – to place a tough force disconcertingly deep behind the enemy lines as the opening move in a major advance. In such instances, however, it is axiomatic that the main force must link up with the airborne elements within a short period of time or the latter, owing to their lack of heavy weaponry, will be crushed (as indeed happened to the British 6th Airborne Division at Arnhem).

XII

An army should have only one line of operations. This should be preserved with care, and only abandoned in the last extremity.

THE line of communication, says Montecuccoli, must be certain and well established, for every army that acts from a distant base and is not careful to keep this line perfectly open, marches upon a precipice. It moves to certain ruin, as may be seen by an infinity of examples. In fact, if the road by which provisions, ammunition, and reinforcements are to be brought up is not entirely secured, if the magazines, the hospitals, the depôts of arms, and the places of supply are not fixed, and commodiously situated, not only the army cannot keep the field, but it will be exposed to the greatest dangers.

CHANDLER: Simplicity is a basic requirement in military plans and always has been. The selection of a single line of operations is obviously advantageous – providing it can be disguised from the enemy for as long as possible. Thus a whole series of diversionary probes would be likely to precede a Soviet main thrust in order to confuse the NATO defence. Clearly, accurate and early divination of the enemy's true intentions is critically important in any such scenario. The problem of using more than one line of operations is that of dissipation of force, and that applies to the concept of probing secondary attacks mentioned above, at least to a degree. As usual, a careful balance has to be struck. But "selection and maintenance of the aim" is one principle of war that surely is mandatory if success is to be achieved.

XIII

The distances which may be allowed between the divisions of an army on the march must depend on the localities or circumstances, and on the object in view.

WHEN an army moves at a distance from the enemy the columns may be disposed along the road so as to favour the artillery and baggage. But when it is marching into action, the different corps must be formed in close columns in order of battle. The generals must take care that the heads of the columns which are to attack together do not outstep each other, and that in approaching the field of action they preserve the relative intervals required for deployment.

The marches that are made preparatory to a battle require, says Frederick, the greatest precaution. With this view he recommends his generals to be particularly on their guard, and to reconnoitre the ground at successive distances, in order to secure the initiative by occupying those positions most calculated to favour an attack. On a retreat, it is the opinion of many generals that an army should concentrate its forces, and march in close columns if it is still strong enough to resume the offensive; for by this means it is easy to form the line when a favourable opportunity presents itself either for holding the enemy in check, or for attacking him if he is not in a situation to accept battle.

Such was Moreau's retreat after the passage of the Adda by the Austro-Russian army. The French general, after having covered the evacuation of Milan, took up a position between the Po and the Tenaro.

His camp rested upon Alexandria and Valentia, two capital fortresses, and had the advantage of covering the roads to Turin and Savona, by which he could effect his retreat in case he was unable to accomplish a junction with the *corps d'armée* of Macdonald, who had been ordered to quit the kingdom of Naples, and hasten his march into Tuscany.

Forced to abandon this position in consequence of the insurrection in Piedmont and Tuscany, Moreau retired upon Asti, where he learned that his communication with the river of Genoa had just been cut off by the capture of Ceva. After several ineffectual attempts to retake this place, he saw that his only safety depended upon throwing himself into the mountains.

To effect this object he directed the whole of his battering-train and heavy baggage by the Col de Fenestrelle upon France, then opening himself a way over the St. Bernard, he gained Scano with his light artillery and the small proportion of field equipment he had been able to preserve.

By this skilful movement he not only retained his communications with France, but was enabled to observe the motions of the army from Naples, and to facilitate his junction with it by directing the whole of his force upon the points necessary for that purpose.

Macdonald, in the meantime, whose only chance of success depended on concentrating his little army, neglected this precaution, and was beaten in three successive actions at the Trebia.

By this retardment of his march he rendered all Moreau's measures to unite the two armies in the plains of the Po useless, and his retreat after his brilliant but fruitless efforts at the Trebia defeated the other dispositions also, which the former had made to come to his support. After all, however, the inactivity of Marshal Souwarrow enabled the French general to accomplish his junction with the remains of the army from Naples. Moreau then concentrated his whole force upon the

Apennines, and placed himself in a situation to defend the important positions of Liguria until the chances of war should afford him an opportunity of resuming the offensive.

When after a decisive battle an army has lost its artillery and equipments, and is consequently no longer in a state to resume the offensive or even to arrest the pursuit of the enemy, it would seem most desirable to divide what remains into several corps, and order them to march by separate and distant routes upon the base of operation and throw themselves into the fortresses. This is the only means of safety, for the enemy, uncertain as to the precise direction taken by the vanquished army, is ignorant in the first instance which corps to pursue, and it is in this moment of indecision that a march is gained upon him. Besides the movements of a small body being so much easier than those of a larger one, these separate lines of march are all in favour of a retreating army.

CHANDLER: Clearly, conditions governing inter-operability of major formations require the most careful consideration. One secret of Napoleon's campaigns was the creation of a "spider's web" of corps marching independently but carefully co-ordinated by *Grand Quartier-Général* on the basis of twice-daily reports received and orders issued. The general rule was that each army corps had to be within one day's forced marching distance (perhaps 20 miles) of one or two others. A corps, as an all-arm formation, could in theory fight several times its own number of opponents without succumbing totally for up to 24 hours – but needed to be reinforced within that time by neighbouring formations. Mutual support and a high degree of flexibility were important principles underlying the "square battalion" (or diamond-shaped marching formation) of strategic manoeuvre as used in 1806 against Prussia. Today, the possibilities of heli-borne or airborne reinforcement of critical areas has added a new dimension to strategic movement and reinforcement capabilities.

Sometimes matters could go awry. Bernadotte's failure to fight at either Jena or Auerstädt on 14 October 1806 and d'Erlon's aimless wandering with I Corps between Quatre Bras and Ligny on 16 June 1815 without actually fighting on either field are two well-known instances when Napoleon's system failed him – largely due to human error. The failure of Horrocks's XXX Corps to link up with the British paratroopers dropped at Arnhem on 21 September 1944, or that of the French relief column sent to break through to Dien Bien Phu in 1954 are two more modern examples. On the other hand, Patton's break-in to relieve the garrison of Bastogne on 25 December 1944 at the height of the battle of the Ardennes, or Slim's fly-in of 5th Indian Division to reinforce beleaguered Imphal in Burma, 1944, are examples of highly successful relief reinforcement opera-tions. Napoleon's maxim remains impeccable.

XIV

Among mountains a great number of positions are always to be found, very strong in themselves, and which it is dangerous to attack. The science of this mode of warfare consists in occupying camps on the flanks, or in the rear of the enemy. . . .

DURING the campaign of 1793, in the Maritime Alps, the French army, under the orders of General Brunet, did all in its power to get possession of the camps at Rans and at Fourches by an attack in front. But these useless efforts served only to increase the courage of the Piedmontese, and to destroy the *élite* of the grenadiers of the republican army. The manœuvres by which Napoleon, without fighting, compelled the enemy to evacuate these positions in 1796, suffice to establish the truth of these principles, and to prove how much success in war depends upon the genius of the general as well as on the courage of the soldier.

CHANDLER: This maxim is in some ways misleading. On the assumption that Napoleon was referring to conventional operations among mountains, it contains a kernel of truth: Clausewitz's dictum that "the defensive is the stronger form of war" is certainly applicable up to a point. But this has to be qualified, for to implement no initiatives means to achieve no results. To lure an enemy from a strong position (as Massena tried unsuccessfully in 1810 when Wellington had occupied the Lines of Torres Vedras, or Hitler's commanders attempted in Yugoslavia, 1942–4, against Tito's partisan army in its mountain fastnesses) it is necessary to feign weakness or to offer some other bait or inducement. In July 1704, Marlborough failed to induce the Elector of Bavaria to leave his fortresses to fight or to negotiate despite the burning by the Allies of some 400 villages. Tactically, too, the maxim needs qualification. It was certainly Wellington's method up to 1812 in the Peninsula boldly to set up a battle situation and then fight essentially defensively (as at Talavera (1809) or Busaco (1810), but it was rarely Napoleon's habit – save possibly at Austerlitz in 1805. Indeed, if one recalls the ill-considered storming of the Pass of Somosierra in late 1808 (when the Emperor threw away brave Polish lives in a pointless series of uphill attacks against Spaniards holding the crest with artillery in position) we may guess that he learned the truth of this dictum the hard way. It is suggested that what Napoleon was really implying was the need to attack from a secure defensive position, just as a modern infantry platoon deploys a fire group or section to cover the line of its attack.

This, however, begs a larger question. In the case of guerrilla warfare, the assailant, *pace* Napoleon, almost invariably enjoys the advantages of being able to select the place, target and time for an attack from protective mountain, forest or jungle lairs. This was as true of the Peninsular War or Tyrolean uprising of 1809 as of Malaya (1948–60) or Vietnam. But, of course, Napoleon never really settled down to deal with irregular forms of warfare beyond defining guerrilla or partisan-type struggles as "wars without a front". Even Clausewitz regarded popular wars as a strategy of the last resort against a conqueror and hence of only residual interest. Only since 1945 has the world come to appreciate the existence of revolutionary wars deliberately fought by low-key methods at the outset to make the most of political factors and the effects of remoseless attrition on the police and military forces supporting the government and society under attack.

XV

The first consideration with a general who offers battle should be the glory and honour of his arms. The safety and preservation of his men is only the second.

In 1645, the French army, under the orders of the Prince of Condé, was on the march to lay siege to Nordlingen, when it was discovered that Count Merci, who commanded the Bavarians, had foreseen this intention, and had entrenched himself in a strong position which defended Nordlingen at the same time that it covered Donawerth.

Notwithstanding the favourable position of the enemy Condé ordered the attack. The combat was terrible. All the infantry in the centre and on the right, after being successively engaged, was routed and dispersed, in spite of the efforts of the cavalry and the reserve, which were likewise carried away with the fugitives. The battle was lost. Condé in despair, having no longer either centre or right to depend upon, directed his march to the left, where Turenne was still engaged. This perseverance reanimated the ardour of the troops. They broke the right wing of the enemy, and Turenne, by a change of front, returned to the attack upon his centre. Night, too, favoured the boldness of Condé. An entire corps of Bavarians, fancying themselves cut off, laid down their arms, and the obstinacy of the French general in this struggle for victory was repaid by possession of the field of battle, together with a great number of prisoners and almost all the enemy's artillery. The Bavarian army beat a retreat, and the next day Nordlingen capitulated.

CHANDLER: As Clausewitz would write later, it is "impossible to make omelettes without breaking eggs". Casualties are inseparable from active prosecution of warfare, but smaller countries such as Great Britain have to be charier about suffering heavy losses than larger continental powers that enjoy superior reservoirs of manpower. Napoleon is thinking here as a continental soldier. Wellington in the Peninsula placed the preservation of England's only field army as his first and overriding priority. Wavell felt much the same in the Middle East, 1940–1, when he was responsible for the fates of nine countries with barely 50,000 British, Commonwealth and Imperial soldiers at his disposal. However, as Wavell demonstrated in his offensives against Italian East Africa and Libya, the advantages of bold attacks – "the glory and honour of his arms" – are not to be discounted as providing the best means to offset daunting numerical disadvantages (in his case over 500,000 Italian troops in the theatre even before the arrival of Rommel in February 1941). Retreats certainly cost an army dear. Moore lost almost half his army retreating to Corunna (1808–9); "General Winter" and the Russians cost Napoleon 80% of his Grand Army that left Moscow in October 1812; the Chinese "Great March" over 6,000 miles to Shenshi Province in the 1930s cost Mao Tse Tung communist lives on the same scale (but those who survived went on to win the Civil War by 1950); Slim's 1000-mile retreat through Burma to north-east India in 1942 had to fight disease, climate and terrain as well as the Japanese, but still the remnants of "Burcorps" found the spirit to cheer their commander at its end. Clearly "brave men" were still "to be found with the colours" in both the last-cited cases. Quality has much more to do with achieving victory than quantity on its own – as the Falklands campaign on land proved.

XVI

It is an approved maxim in war never to do what the enemy wishes you to do, for this reason alone, because he wishes *it. A field of battle, therefore, which he has previously studied and reconnoitred should be avoided.* . . .

IT was without due regard to this principle that Marshal Villeroi, on assuming the command of the army of Italy during the campaign of 1701, attacked with unwarrantable presumption Prince Eugene of Savoy in his entrenched position of Chiari on the Oglio. The French generals, Catinat among the rest, considered the post unassailable, but Villeroi insisted, and the result of this otherwise unimportant battle was the loss of the *élite* of the French army. It would have been greater still but for Catinat's exertions.

It was by neglecting the same principle that the Prince of Condé, in the campaign of 1644, failed in all his attacks upon the entrenched position of the Bavarian army. The Count Merci, who commanded the latter, had drawn up his cavalry skilfully upon the plain, resting upon Freyberg, while his infantry occupied the mountain. After many fruitless attempts the Prince of Condé, seeing the impossibility of dislodging the enemy, began to menace his communications, but the moment Merci perceived this he broke up his camp and retired beyond the Black Mountains.

CHANDLER: As von Moltke the Elder remarked, the enemy can usually be expected to pursue the seventh course of action when one has prepared for only six possibilities; such is the nature of "friction" in war. Napoleon, probably unwittingly, broke his own maxim by attacking Wellington on the Waterloo position, as the Duke had reconnoitred it in 1814. As for the lack of wisdom in attacking a fortified position, we have already cited the case of Somosierra in 1808. Seventy years later, the Russians learned much the same lesson against the strong Turkish defences of Plevna in the Balkans. Another notable instance was the rash German offensive against the prepared Kursk salient in Russia (1943) – a "set up" situation if ever there was one. Undoubtedly, therefore, the rule generally holds good: one thinks of the German avoidance of the main Maginot Line by means of a turning action through the Ardennes in 1940, or Massena's frustration by the Lines of Torres Vedras in 1810 (which by their nature permitted no such manœuvre). Envelopment attacks, or "the indirect approach", have much to offer, as Napoleon showed time and again between 1796 and 1814.

Equally, however, direct attacks on strongly-fortified areas can be successful, if almost invariably costly. General Giap failed against Da Nang in 1967 although he succeeded against Dien Bien Phu in 1954. On the other hand, Marlborough successfully stormed for minimal loss the Lines of Brabant in 1705 and those of *Ne Plus Ultra* in 1711 by sleight of hand and superb deception, which in each case caught the French off balance at the critical point and time.

XVII

In a war of march and manœuvre, if you would avoid a battle with a superior army, it is necessary to entrench every night, and occupy a good defensive position. . . .

THE campaign of the French and Spanish army, commanded by the Duke of Berwick, against the Portuguese, in the year 1706, affords a good lesson on this subject. The two armies made almost the tour of Spain. They began the campaign near Badajoz, and after manœuvring across both Castiles, finished it in the kingdom of Valencia and Murcia. The Duke of Berwick encamped his army eighty-five times, and although the campaign passed without a general action, he took about ten thousand prisoners from the enemy. Marshal Turenne also made a fine campaign of manœuvre against the Count Montecuccoli in 1675.

The Imperial army having made its dispositions to pass the Rhine at Strasburg, Turenne used all diligence, and throwing a bridge over the river near the village of Altenheim, three leagues below Strasburg, he crossed with the French army, and encamped close to the little town of Velstet, which he occupied. This position covered the bridge of Strasburg, so that by this manœuvre Turenne deprived the enemy of all approach to that city.

Upon this Montecuccoli made a movement with his whole army, threatening the bridge at Altenheim, by which the French received their provisions from Upper Alsace.

As soon as Turenne discovered the design of the enemy, he made a rapid march with his whole force upon the village of Altenheim. This intermediate position between the two bridges, which he wished to preserve, gave him the advantage of being able to succour either of these posts before the enemy had time to carry them. Montecuccoli, seeing that any successful attack upon the bridges was not to be expected, resolved to pass the Rhine below Strasburg, and with this view returned to his first position at Offenburg. Marshal Turenne, who followed all the movements of the Austrian army, brought back his army also to Velstet.

In the meantime, this attempt of the enemy having convinced the French general of the danger to which his bridge had exposed him, removed it nearer to that of Strasburg, in order to diminish the extent of ground he had to defend. Montecuccoli having commanded the magistrates to collect materials for a bridge, moved to Scherzheim to receive them, but Turenne again defeated his projects by taking a position at Freistett, where he occupied the islands of the Rhine and immediately constructed a stockade.

Thus it was that, during the whole of this campaign, Turenne succeeded in gaining the initiative of the enemy, and obliging him to follow his movements. He succeeded also by a rapid march in cutting off Montecuccoli from the town of Offenburg, from whence he drew his supplies; and would, no doubt, have prevented the Austrian general from effecting his junction with the corps of Caprara, had not a cannon-shot terminated this great man's life.

CHANDLER: Modern wars, the struggle in the Gulf apart, tend to be full of manœuvre and movement in the age of aircraft, the helicopter and the tank. Such largely static struggles as the Western Front from 1915 to 1918 are now exceptional. Fluid operations consequently demand much attention to security, and that includes the creation of good all-round defensive positions each night or whenever a force halts for more than a brief pause. To fail in this is to invite enemy attack, particularly by a stronger opponent. Even when a strong natural position is available, its advantages need reinforcing by preparing minefields, wire obstacles and weapon pits to make the most of it. The Roman army invariably built itself an entrenched and palisaded camp each night while in hostile territory, every legionary being expected to carry two stakes and a spade or pick in the same way as the modern infantryman carries an entrenching tool. Digging positions at the close of a hard day is never popular, but it is an inescapable security precaution on active service.

XVIII

A general of ordinary talent occupying a bad position, and surprised by a superior force, seeks his safety in retreat; but a great captain supplies all deficiencies by his courage, and imposing upon the enemy with a confident front, marches boldly to battle. . . .

IN 1653 Marshal Turenne was surprised by the Prince of Condé, in a position in which his army was completely compromised. He had the power, indeed, by an immediate retreat, of covering himself by the Somme, which he possessed the means of crossing at Peronne, and from whence he was distant only half a league; but fearing the influence of this retrograde movement on the *moral* of his army, Turenne balanced all disadvantages by his courage, and marched boldly to meet the enemy with very inferior forces. After marching a league he found an advantageous position, where he made every disposition for a battle. It was three o'clock in the afternoon, but the Spaniards, exhausted with fatigue, hesitated to attack him, and Turenne having covered himself with entrenchments during the night, the enemy no longer dared to risk a general action, and broke up his camp.

CHANDLER: Mediocre generals holding poor positions and seeking refuge in retreat before even inferior forces, never mind superior ones, are not hard to discover in 20th century military history. One thinks of such Italian commanders as Generals Maletta and "Electric-Whiskers" Berghenzoli being forced out of position after position in Egypt and then Libya by General O'Connor's Western Desert Force in 1940–1, which never numbered more than 20,000 men. Similarly, the British defence of the Malayan Peninsula against the smaller Japanese army of General Mutaguchi in late 1941 to early 1942 left something to be desired, although Japanese air superiority and continuous enveloping marches through "impenetrable" jungle had much to do with the incessant British retreats over some 500 miles down to Singapore island.

History also provides many instances of great captains overcoming deficiencies by bold, determined action – or sheer bluff. Marlborough, rarely at a numerical advantage before 1709, was a master at imposing his will upon stronger foes. The risk-defying 300-mile march to the Danube from The Netherlands in 1704 is a good strategic example; his forcing battles on Tallard at Blenheim (1704), on Villeroi at Ramillies (1706) and on Vendôme at Oudenarde (1708) were all masterpieces of operational art. Similarly, Frederick the Great's bold use of the central position in 1757 to push back one encircling foe after another (resulting in the great victories of Rossbach and Leuthen) is another. Napoleon's record on this score in both 1796 and 1815 – at each extremity of his active career – needs no elaboration here. He was a master at seizing and exploiting the initiative against superior foes, at mentally dominating his opponent, even if on some occasions his opportunistic gambling failed to come off (as at Waterloo). Rommel and Giap were generals made in the same mould in at least this particular respect.

XIX

The transition from the defensive *to the* offensive *is one of the most delicate operations in war.*

IT is by studying the first campaign of Napoleon, in Italy, that we learn what genius and boldness may effect in passing with an army from the *defensive* to the *offensive*. The army of the allies, commanded by General Beaulieu, was provided with every means that could render it formidable. Its force amounted to eighty thousand men, and two hundred pieces of cannon. The French army, on the contrary, could number scarcely thirty thousand men under arms, and thirty pieces of cannon. For some time there had been no issue of meat, and even the bread was irregularly supplied.

The infantry was ill-clothed, the cavalry wretchedly mounted, all the draft-horses had perished from want, so that the service of the artillery was performed by mules. To remedy these evils, large disbursements were necessary, and such was the state of the finances, that the government had only been able to furnish two thousand louis for the opening of the campaign. The French army could not possibly exist in this state. To advance or to retreat was absolutely necessary. Aware of the advantage of surprising the enemy at the very outset of the campaign by some decisive blow, Napoleon prepared for it by recasting the *moral* of his army.

In a proclamation full of energy, he reminded them that an ignoble death alone remained for them if they continued on the defensive; that they had nothing to expect from France, but everything to hope from victory. "Abundance courts you in the fertile plains of Italy," said he; "are you deficient, soldiers, in constancy or in courage?"

Profiting by the moment of enthusiasm which he had inspired, Napoleon concentrated his forces in order to fall with his whole weight on the different corps of the enemy. Immediately afterwards the battles of Montenotte, Milesimo, and Mondovi added fresh confidence to the high opinion already entertained by the soldier for his chief, and that army which only a few days ago was encamped amid barren rocks, and consumed by famine, already aspired to the conquest of Italy. In one month after the opening of the campaign, Napoleon had terminated the war with the King of Sardinia, and conquered the Milanese. Rich cantonments soon dispelled from the recollection of the French soldier the misery and fatigue attendant upon this rapid march, while a vigilant administration of the resources of the country reorganised the *matériel* of the French army, and created the means necessary for the attainment of future success.

CHANDLER: History shows that this observation is true indeed. Here the requirements of military art call for careful calculation and a sense of timing if success is to be achieved. What is possible at one moment may not be feasible at the next; opportunity is a fleeting matter, and it requires a good commander to recognise it and seize it. The timing of Kellermann the Younger's charge at the critical moment at Marengo (1800), of the precise moment for unleashing Soult's two divisions against the Pratzen Heights from the mist-enshrouded valley at Austerlitz (1805), or Wellington's timing of Pakenham's attack against the advance-guard of Marmont's rashly over-extended army at Salamanca (1812) – are obvious examples at the tactical and operational levels of war. In the First World War, Generals French and Haig all too often got their timings wrong in launching their notoriously expensive offensives on the Western Front at Ypres and the Somme. In the Second World War, Rommel was famed for judging his moments to attack with skill, as was, on a far larger scale, Marshal Zhukov on the Eastern Front from 1943 onwards. Israeli skills in matters of timing have also been notable.

XX

It may be laid down as a principle that the line of operation should not be abandoned. But it is one of the most skilful manœuvres in war to know how to change it, when circumstances authorise or render this necessary. . . .

FREDERICK sometimes changed his line of operation in the middle of a campaign; but he was enabled to do this because he was manœuvring at that time in the centre of Germany, an abundant country, capable of supplying all the wants of his army in case his communications with Prussia were intercepted.

Marshal Turenne, in the campaign of 1674, gave up his line of communication to the allies in the same manner, but like Frederick he was carrying on the war at this time in the centre of Germany, and having fallen with his whole forces upon Rain, he took the precaution of securing to himself a depôt upon which to establish his base of operation. By a series of manœuvres, marked alike by audacity and genius, he subsequently compelled the imperial army to abandon its magazines, and retire into Austria for winter quarters.

But these are examples which it appears to me should only be imitated when we have taken full measure of the capacity of our adversary, and above all, when we see no reason to apprehend an insurrection in the country to which we transfer the theatre of war.

CHANDLER: Again, skill in this kind of decision picks out the great commander from the mediocre. Marlborough, in 1704, twice changed his line of operations to fool the foe as to his real intentions, advancing first on the Moselle and later towards Strasbourg to conceal his real intention of marching to the distant Danubian theatre of war. He also switched his lines of communication from the Rhine to the River Main at a critical point so as to safeguard them. Napoleon achieved this in 1805 before Austerlitz and next year before Jena. In modern times one thinks of General Giap using the Khe Sanh offensive and then his offer of an armistice in late 1967 to conceal preparations for his main blow – the Tet Offensive of late January 1968, when 100 South Vietnamese towns and cities were almost simultaneously attacked during the cease-fire.

XXI

When an army carries with it a battering-train, or large convoys of sick and wounded, it cannot march by too short a line upon its depôts.

IT is above all in mountainous countries, and in those interspersed with woods and marshes, that it is of importance to observe this maxim, for the convoys and means of transport being frequently embarrassed in defiles, an enemy by manœuvring may easily disperse the escorts, or make even a successful attack upon the whole army when it is obliged, from the nature of the country, to march in an extended column.

CHANDLER: This sensible security precaution is designed to safeguard large convoys from the danger of interdictive attack. The closer such columns are kept to the forward depots, established progressively as an army advances, the safer they will be. Conversely, the more tortuous or indirect the line of operations being followed the greater is the danger of their being intercepted and taken out by the enemy.

William III lost much of his siege train to such a raid by Patrick Sarsfeld during his campaign against James II in Ireland (1690). A generation later, it was the almost total loss of General Lewenhaupt's reinforcement convoy (1708) that doomed Charles XII and his Swedish army to ultimate defeat in the depths of Peter the Great's Russia the following year. In both these cases it can be argued that the targets were allowed to become isolated from close support – depots or nearby troops on land. Logistic considerations loom large in the minds of military men. "For God's sake do not risk the cannon" was Marlborough's heartfelt and unusually blunt appeal to Cadogan as the Great Train of siege materials – taking up 30 miles of road – moved ponderously towards Lille (1708). The loss of field ambulances to the enemy has always been bad for morale – not least when fighting the Japanese in the Second World War.

The modern application of this maxim is clear: the special convoys carrying Cruise missiles from their bases to deployment areas are potential targets for disruptive activities and even sabotage by extremist groups, while the deployment of Pershings, SS-20s, and other highly mobile nuclear weaponry in time of transition to war poses anxieties to their modern owners similar to those expressed by Marlborough in 1708. The maxim enjoins staying close to one's defended areas.

XXII

The art of encamping in a position is the same as forming the line for battle in that position. To this end the artillery should be advantageously placed. . . .

FREDERICK has remarked that in order to be assured that your camp is well placed you should see if by making a small movement you can oblige the enemy to make a greater, or if, after having forced him to retrograde one march, you can compel him to fall back another.

In defensive war all camps should be entrenched in the front and wings of the position they occupy, and care should be taken that the rear is left perfectly open. If you are threatened with being turned, arrangements should be made beforehand for taking up a more distant position, and you should profit by any disorder in the enemy's line of march to make an attempt upon his artillery or baggage.

CHANDLER: It is today axiomatic that armies engaged in actual warfare or serious manœuvres take up carefully selected positions capable of all-round defence and relevant to their anticipated battle roles when halting whether for a long or short period of time. In the 17th and 18th centuries, armies always encamped in battle order, ready for an alarm. Thus at Sedgemoor (1685), Feversham's infantry was able to take up its battle stations to the west of its tent lines near Westonzoyland with minimum confusion or delay to meet Monmouth's rash night attack. The Royal Army's position was also well-chosen behind the Bussex rhyne or large drainage ditch so as to avoid being outflanked. In 1800, it was the over-extended camping position taken up by the Army of the Reserve on the night of 13 June that compromised Napoleon himself during the morning and afternoon battle of Marengo (1800). Only Desaix's timely arrival in the evening saved the day. The point about artillery is of interest. At Sedgemoor the Royal artillery was not well positioned to protect the position from the line of attack Monmouth deliberately chose so as to avoid the guns. They were simply massed on the left to cover the obvious approach – the Bridgwater road.

In late 1953, the design of the French fortified base deep in enemy territory at Dien Bien Phu was also flawed in a similar way. All artillery was massed in an under-protected area to the south of the main positions, and it proved incapable of taking on the Vietminh guns sited in the surrounding mountains because French intelligence had critically under-estimated the calibre of an impor-tant part of General Giap's available artillery. The French disaster that ensued in the spring of 1954 was the result.

XXIII

When you are occupying a position which the enemy threatens to surround, collect all your strength immediately, and menace him with an offensive movement. . . .

THIS was the manœuvre practised by General Desaix in 1798 near Radstadt. He made up for inferiority in numbers by audacity, and maintained himself the whole day in position in spite of the vigorous attacks of the Archduke Charles. At night he effected his retreat in good order, and took up a position in the rear.

It was in accordance also with this principle, in the same campaign, that General Moreau gave battle at Biberach to secure his retreat by the passes of the Black Mountains. A few days after he fought at Schliengen with the same object. Placed in a good defensive position, he menaced the Archduke Charles by a sudden return to the offensive, while his artillery and baggage were passing the Rhine by the bridge of Huningen, and he was making all the necessary dispositions for retiring behind that river himself.

Here, however, I would observe that the execution of such offensive demonstrations should be deferred always till toward the evening, in order that you may not be compromised by engaging too early in a combat which you cannot long maintain with success.

Night, and the uncertainty of the enemy after an affair of this kind, will always favour your retreat if it is judged necessary; but with a view to mask the operation more effectually, fires should be lighted all along the lines to deceive the enemy, and prevent him from discovering this retrograde movement, for in a retreat it is a great advantage to gain a march upon your adversary.

CHANDLER: This situation is well illustrated by General Slim's preparations for the battle of Kohima-Imphal in Burma, 1944. At both positions a tight perimeter was established for all-round defence, and outpost actions well to the south and east on the Tiddim road and elsewhere seriously held up the Japanese offensive. Ten years later, it was the Colonel de Castries' refusal to undertake aggressive patrolling in the "impassable" hills surrounding Dien Bien Phu that surrendered all the initiative to General Giap. There had been a similar lack of local initiative by General von Paulus during the early stages of the Russian siege of Stalingrad in late 1942; he called off his one major break-out attempt to conserve petrol supplies when his forces were, in fact, within sight of complete success. The outcome was identical: neither the eleventh-hour promotion of de Castries to general officer by General Navarre or of von Paulus to field marshal by Adolf Hitler prevented the total destruction of the élite French garrison or the German Sixth Army respectively.

In 1815 at Waterloo, Wellington deliberately used the outposts of Hougoumont, La Haie Sainte and La Haie–Fischermont to hold up the development of Napoleon's attack, and to help guard against the possibility of local outflanking attack (while a force held around Hal to the west secured his position from being strategically turned). As it transpired, the battle was very much "a near-run thing". "Give me night or give me Blücher" was the Duke's prayer at about 5.45 p.m. on 18 June.

XXIV

Never lose sight of this maxim, that you should establish your cantonments at the most distant and best protected point from the enemy, especially when a surprise is possible.

In the campaign of 1645 Marshal Turenne lost the battle of Marienthal by neglecting this principle, for if, instead of reassembling his divisions at Erbsthausen, he had rallied his troops at Mergentheim behind the Tauber, his army would have been much sooner reunited, and Count Merci, in place of finding only three thousand men to fight at Erbsthausen (of which he was well informed), would have had the whole French army to attack in a position covered by a river.

Some one having indiscreetly asked Viscount Turenne how he had lost the battle of Marienthal – "By my own fault," replied the marshal; "but," he added, "when a man has committed no faults in war, he can only have made it a short time."

CHANDLER: Clearly, to camp within close range of an enemy is to run the risk of being subjected to a surprise attack. It was for this reason that Feversham selected so sensible a position before Sedgemoor (see above, XXII), and further strengthened it with well-placed outposts placed to the fore – one of which, near Chedzoy, successfully detected Monmouth's crossing of the Langmoor rhyne despite the thick fog prevailing at the time. The alarm then raised at about 2 a.m. permitted the Royal Army to stand-to in its pre-designated alarm posts behind the Bussex rhyne before the rebel army could come to grips. The advantages of surprise and numbers were thus lost to the attacker. By comparison, it was Blücher's exposed positioning of part of his army near Charleroi and Fleurus in June 1815 that laid him open to Napoleon's surprise offensive on 15 June and further compelled the Prussians to fight the French on an unsuitable battlefield around Ligny the next afternoon – where they took a hard drubbing as a result. Similarly, Marshal

Graziani's inadequate security measures in the scattered Nibeiwa and Sofafi camps over a 50 mile area near Sidi Barrani, greatly assisted General O'Connor's surprise attack with inferior forces in western Egypt on 9 December 1940.

XXV

When two armies are in order of battle, and the one has to retire over a bridge while the other has the circumference of the circle open, all the advantages are in favour of the latter. . . .

THIS was the position of the French army at the famous battle of Leipzig, which terminated the campaign of 1813 so fatally for Napoleon; for the battle of Hanau was of no consequence comparatively in the desperate situation of that army.

It strikes me that in a situation like that of the French army previous to the battle of Leipzig, a general should never calculate upon any of those lucky chances which may arise out of a return to the offensive, but that he should rather adopt every possible means to secure his retreat. With this view he should immediately cover himself with good entrenchments to enable him to repel with inferior numbers the attack of the enemy, while his own equipments are crossing the river. As fast as the troops reach the other side, they should occupy positions to protect the passage of the rear-guard, and this last should be covered by a *tête de pont* as soon as the army breaks up its camp. During the wars of the Revolution too little regard was paid to entrenchments, and it is for this reason we have seen large armies dispersed after a single reverse, and the fate of nations compromised by the issue of one battle.

CHANDLER: The French quandary at the battle of the Beresina in November 1812 is as good an example as that of Leipzig (1813) cited by the earlier editor. The Russians could bring fire upon the two rickety bridges and also certainly tried a major outflanking attack from the south over the Beresina, but Napoleon was able to fight off Admiral Tshitsagov's onslaught and complete the evacuation of his main body. He lost many stragglers, however, when the bridges had to be destroyed.

In 1942 General Percival successfully withdrew his forces over the causeway into Singapore island – but did not long forestall the Japanese crossing of the Johore Strait. On 8 March 1945 it was just such a bold attack as Napoleon enjoins – together with faulty German demolition charges – that enabled a small American force under Lieutenant Carl Timmerman to capture intact the Rhine bridge at Remagen, south of Bonn. Today, of course, such a tempting bottleneck would invite air or tactical nuclear attack.

XXVI

It is contrary to all principle to make corps which have no communication act separately, against a central force whose communications are open.

THE Austrians lost the battle of Hohenlinden by neglecting this principle. The imperial army, under the orders of the Archduke John, was divided into four columns, which had to march through an immense forest, previous to their junction in the plain of Anzing, where they intended to surprise the French. But these different corps, having no direct communication, found themselves compelled to engage separately with an enemy who had taken the precaution of concentrating his masses, and who could move them with facility in a country with which he had been long previously acquainted.

Thus the Austrian army, enclosed in the defiles of the forest with its whole train of artillery and baggage, was attacked in its flanks and rear, and the Archduke John was only enabled to rally his dispersed and shattered divisions under cover of the night. The trophies obtained by the French army on this day were immense.

They consisted of eleven thousand prisoners, one hundred pieces of cannon, several stand of colours, and all the baggage of the enemy.

The battle of Hohenlinden decided the fate of the campaign of 1800, and his brilliant and well-merited success placed Moreau in the rank of the first general of the age.

CHANDLER: This maxim highlights the problems faced by forces driven apart by the application of Napoleon's strategy of the "central position" – as he achieved in his first operation as a commanding general against the Austrians at Montenotte (1796) and again at the outset of his last operation against Wellington and Blücher at Charleroi on 15 June 1815. Such forces acting independently without secure links between them are

courting defeat in detail. This was well illustrated by the
fate of the divided Russian First and Second Armies at
Tannenberg in late August 1914 at the hands of
Hindenburg's smaller but centrally-positioned and well-
integrated German Eighth Army. "Fight divided, hang
together."

XXVII

*When an army is driven from a position, the retreating columns
should rally always sufficiently in the rear to prevent the enemy's
interference with this object. . . .*

ONE great advantage which results from rallying your
columns on a point far removed from the field of battle,
or from the position previously occupied, is that the
enemy is left in uncertainty of the direction you mean to
take.

If he divides his force to pursue you he exposes himself
to see his detachments beaten in detail, especially if you
have exerted all due diligence, and have effected the
junction of your troops in sufficient time to get between
his columns and disperse them one after the other.

It was by a manœuvre of this kind in the campaign of
Italy, in 1799, that General Melas gained the battle of
Genola.

General Championet commanded the French army,
and endeavoured to cut off the communication of the
Austrians with Turin, by employing troops which
manœuvred separately to get into their rear. Melas, who
divined his project, made a retrograde march by which
he persuaded his adversary he was in full retreat,

although the real object of his movement was to concentrate his forces at the point fixed for the junction of the different detachments of the French army, and which he beat and dispersed one after another by his great superiority in numbers. The result of this manœuvre, in which the Austrian general displayed vigour, decision, and *coup d'œil*, secured to him the peaceable possession of Piedmont.

It was also by the neglect of this principle that General Beaulieu, who commanded the Austro-Sardinian army in the campaign of 1796, lost the battle of Milesimo after that of Montenotte.

His object, in endeavouring to rally his different corps upon Milesimo, was to cover the high roads of Turin and Milan; but Napoleon, aware of the advantages arising from the ardour of troops emboldened by recent success, attacked him before he could assemble his divisions, and by a series of skilful manœuvres succeeded in separating the combined armies. They retired in the greatest disorder – the one by the road of Milan, the other by that of Turin.

CHANDLER: A good example of the proper observance of this maxim is the conduct of the Prussian army immediately after their considerable defeat at Ligny on 16 June 1815. Instead of fleeing towards their base at Namur, General Gneisenau (in Blücher's enforced absence), order an initial retreat to regroup at Wavre, 10 miles to the North. This decision not only fooled the French on the 17th but also had an important bearing on the outcome at Waterloo on the 18th as it permitted Blücher to move three corps to Wellington's assistance at Waterloo, leaving just one to hold off Grouchy's belated pursuit. It was lack of observance of this maxim that greatly increased the scale of the Turkish defeat at Meggido in Palestine (1918). Following their heavy defeat by Allenby's army, their retreating columns were caught by British aircraft in the defiles leading down to the Jordan, and hard hit. This also was the fate of the Egyptian Sinai army in June 1967, when they were caught retreating through the Mitla Pass by the unchallenged Israeli Air Force and General Tal's advancing armour, and pulverised.

XXVIII

No force should be detached on the eve of a battle, because affairs may change during the night, either by the retreat of the enemy, or by the arrival of large reinforcements which might enable him to resume the offensive, and render your previous dispositions disastrous.

IN 1796 the army of the Sambre and the Meuse, commanded by General Jourdan, effected a retreat which was rendered still more difficult by the loss of his line of communication. Seeing, however, the forces of the Archduke Charles disseminated, Jourdan, in order to accomplish his retreat upon Frankfort, resolved to open himself a way by Wurtzburg, where there were at that moment only two divisions of the Austrian army. This movement would have been attended with success, if the French general, believing he had simply these two divisions to contend with, had not committed the error of separating himself from the corps of Le Fèvre, which he left at Schweinfurt to cover the only direct communication of the army with its base of operation.

The commission of this fault at the outset, added to some slowness in the march of the French general, secured the victory to the Archduke, who hastened to concentrate his forces.

The arrival of the two divisions also of Kray and Wartesleben during the battle, enabled him to oppose fifty thousand men to the French army, which scarcely numbered thirty thousand combatants. This last was consequently beaten and obliged to continue its retreat by the mountains of Fuldes, where the badness of the roads could be equalled only by the difficulty of the country.

The division of Le Fèvre, amounting to fourteen thousand men, would, in all probability, have turned the scale in favour of Jourdan, had this last not unfortunately conceived that two divisions only were opposing his passage to Wurtzburg.

CHANDLER: Another case of "the biter bit". Napoleon committed exactly this error on 13 June 1800 when he detached first Lapoype to march northwards towards the River Po and then Desaix at the head of Boudet's division to march southwards towards Novi. Having thus reduced his fighting strength by almost 50%, the Army of the Reserve (23,000 strong) found itself attacked early the next morning by General Melas (27,000) from the fortress of Alessandria over the Fontanove brook. "For God's sake come up if you still can," was

Napoleon's urgent order of recall to Desaix. Fortunately that officer was already returning, "marching on the sound of the cannon", and the arrival of Boudet's 5,300 men in the late afternoon permitted a lost battle to be regained.

XXIX

When you have resolved to fight a battle, collect your whole strength. Dispense with nothing. A single battalion sometimes decides the day.

I THINK it here desirable to observe that it is prudent before a battle to fix upon some point in rear of the reserve for the junction of the different detachments; for, if from unforeseen circumstances these detachments should be prevented from joining before the action has commenced, they would be exposed, in case a retrograde movement should have been found necessary, to the masses of the enemy. It is desirable also to keep the enemy in ignorance of these reinforcements, in order to employ them with greater effect. A seasonable reinforcement, says Frederick, renders the success of a battle certain, because the enemy will always imagine it stronger than it is, and lose courage accordingly.

CHANDLER: "It is one drop that causes the bucket to run over," commented Napoleon. Clearly the principle of concentration of force for battle is important – but that of economy of force has also to be kept in mind. Too many troops present at one scene of action may result in too few being available to deal with crises elsewhere. Today, of course, the danger of attracting major air attack or even a tactical nuclear weapon strike makes the matter of "concentration" even more complicated than formerly. The damage wrought on von Schweppenburg's Panzer Group Headquarters and divisions by Allied fighter-bomber and rocket-firing aircraft around La Panne on 8 June 1944 as they prepared to attack the D-Day beach-head is a good example. Nevertheless, the adage that one can never be too strong at a given point in war holds much validity.

XXX

Nothing is so rash and so contrary to principle, as to make a flank march before an army in position, especially when this army occupies heights at the foot of which you are forced to defile.

It was by a neglect of this principle that Frederick was beaten at Kolin in the first campaign of 1737. Notwithstanding prodigies of valour, the Prussians lost fifteen thousand men and a great portion of their artillery, while the loss of the Austrians did not exceed five thousand men. The consequence of this battle was more unfortunate still, since it obliged the King of Prussia to raise the siege of Prague and to evacuate Bohemia.

It was also by making a flank march before the Prussian army that the French lost the disgraceful battle of Rosbach.

This imprudent movement was still more to be reprehended, because the Prince de Soubise, who commanded the French army, had carried his indiscretion so far, as to manœuvre, without either advanced guards or flanking corps, in presence of the enemy. The result was that this army, consisting of fifty thousand men, was beaten by six battalions and thirty squadrons. The French lost seven thousand men, twenty-seven standards, and a great number of cannon.

The Prussians had only three hundred men *hors de combat*.

Thus, by having forgotten this principle, *that a flank march is never to be made before an enemy in line of battle*, Frederick lost his army at Kolin, and Soubise at Rosbach lost both his army and his honour.

CHANDLER: This was precisely the position the Duke of Marlborough manœuvred Marshal Boufflers into at the end of the "Heaths of Peer" operation in early August 1702. Trapped between Marlborough's army holding high ground and the fortress of Maastricht, and forced to escape between them, the French appeared doomed to defeat. However, the Dutch deputies attached to Marlborough's army vetoed action as unnecessary against a foe already in full retreat. "This was very fortunate for us," commented Marshal Berwick, a French participant. In 1704, however, Marlborough conducted a dangerous flank march down the east bank of the Rhine from Coblenz to near Phillipsburg on his way to the Danube theatre, despite the presence of Marshal Villeroi's and General de Coigny's far stronger forces on the west bank. Extreme daring, good security, brilliant administration and, above all, clever bluffs brought complete success to this bold manœuvre.

In more recent times, the Israeli army several times courted a similar risk in 1967 during the desperate fighting for the Syrian-held Golan Heights. Once again, the combination of necessity and daring brought success to their arms.

XXXI

When you determine to fight a great battle, reserve to yourself every possible chance of success, more particularly if you have to deal with an adversary of superior talent; for if you are beaten, even in the midst of your magazines and your communications, "woe to the vanquished!"

WE should make war, says Marshal Saxe, without leaving anything to hazard, and in this especially consists the talent of a general. But when we have incurred the risk of a battle, we should know how to profit by the victory, and not merely content ourselves according to custom with possession of the field.

It was by neglecting to follow up the first success that the Austrian army, after gaining the field of Marengo, saw itself compelled on the following day to evacuate the whole of Italy.

General Melas, observing the French in retreat, left the direction of the movements of his army to the chief of his staff, and retired to Alexandria to repose from the fatigues of the day. Colonel Zach, equally convinced with his general that the French army was completely broken, and consisted only of fugitives, formed the divisions in column of route.

By this arrangement the imperial army prepared to enter upon its victorious march in a formation not less than three miles in depth.

It was near four o'clock when General Desaix rejoined

the French army with his division. His presence restored in some degree an equality between the contending forces; and yet Napoleon hesitated for a moment whether to resume the offensive, or to make use of this corps to secure his retreat. The ardour of the troops to return to the charge decided his irresolution. He rode rapidly along the front of his divisions, and addressing the soldiers, "*We have retired far enough for today,*" said he; "*you know I always sleep upon the field of battle.*"

The army, with unanimous shout, proclaimed to him a promise of victory. Napoleon resumed the offensive. The Austrian advanced guard, panic-struck at the sight of a formidable and unbroken body presenting itself suddenly at a point where a few moments before only fugitives were to be seen, went to the right about, and carried disorder into the mass of its columns. Attacked immediately afterwards with impetuosity in its front and flanks, the Austrian army was completely routed.

Marshal Daun experienced nearly the same fate as General Melas at the battle of Torgau in the campaign of 1760.

The position of the Austrian army was excellent. It had its left upon Torgau; its right on the plateau of Siptitz, and its front covered by a large sheet of water.

Frederick proposed to turn its right in order to make an attack upon the rear. For this purpose he divided his army into two corps, the one under the orders of Ziethen, with instructions to attack in front, following the edge of the water; the other under his own immediate command, with which he set out to turn the right of the Austrians; but Marshal Daun, having had intimation of the movements of the enemy, changed his front by countermarch-ing, and was thus enabled to repel the attacks of Frederick, whom he obliged to retreat. The two corps of the Prussian army had been acting without communication. Ziethen, in the meantime, hearing the fire recede, concluded that the king had been beaten, and commenced a movement by his left in order to rejoin him; but falling in with two battalions of the reverse, the Prussian general profited by this reinforcement to resume the offensive. Accordingly, he renewed the attack with vigour, got possession of the plateau of Siptitz, and soon after of the whole field of battle. The sun had already set when the King of Prussia received the news of this unexpected good fortune. He returned in all haste, took advantage of the night to restore order in his disorganised army, and the day after the battle occupied Torgau.

Marshal Daun was receiving congratulations upon his victory, when he heard that the Prussians had resumed the offensive. He immediately commanded a retreat, and at daybreak the Austrians repassed the Elbe with the loss of twelve thousand men, eight thousand prisoners, and forty-five pieces of cannon.

After the battle of Marengo, General Melas, although in the midst of his fortresses and magazines, saw himself compelled to abandon everything in order to save the wreck of his army.

General Mack capitulated after the battle of Ulm, although in the centre of his own country.

The Prussians, in spite of their depôts and reserves, were obliged, after the battle of Jena, and the French after that of Waterloo, to lay down their arms.

Hence, we may conclude, that the misfortune that

results from the loss of a battle, does not consist so much in the destruction of men and of *matériel* as in the discouragement which follows this disaster. The courage and confidence of the victors augment in proportion as those of the vanquished diminish; and whatever may be the resources of an army, it will be found that a retreat will degenerate rapidly into a rout, unless the general-in-chief shall succeed, by combining boldness with skill, and perseverance with firmness, in restoring the *moral* of his army.

CHANDLER: There will never be an occasion when success in battle can be absolutely assured as so many variables come into play, but it is obviously the sense of this maxim that action should not be rushed into in any haphazard fashion. Risk is always involved, but it should be carefully calculated. A good example of the penalties involved in fighting a superior opponent "even in the midst of your magazines and your communications" is again the Falklands Campaign of 1982. The Galtieri regime, having seized the islands on 2 April, certainly built up large magazines of munitions and supplies over the next month, and also enjoyed a far shorter line of communication back to Argentina than the British 8,000-mile system. But General Menendez and his strong garrison (mainly conscripts) were quite outclassed by the 5,000 British regular Royal Marines, Paratroopers, Guardsmen and Ghurkas that were successfully landed on the east Island by late May. Defeated in a series of brisk actions, Menendez had no option but to surrender unconditionally on 14 June. Galtieri had

underestimated British and UNO reaction, miscalculated the state of British capability to mount a rapid and
effective military response, and relied upon the onset of
the South Atlantic winter to earn Argentina at least a six-
month respite. On all three counts he was proved wrong.
Vae victis!

XXXII

*The duty of an advanced guard does not consist in advancing or
retiring, but in manœuvring. An advanced guard should be composed of light cavalry, supported by a reserve of heavy, and by
battalions of infantry, supported also by artillery. . . .*

IT was the opinion of Frederick that an advanced guard
should be composed of detachments of troops of all arms.
The commander should possess skill in the choice of
ground and he should take care to be instantly informed,
by means of numerous patrols, of everything passing in
the enemy's camp.

In war, it is not the business of an advanced guard to
fight, but to observe the enemy, in order to cover the
movements of the army. When in pursuit the advanced
guard should charge with vigour, and cut off the baggage
and insulated corps of the retiring enemy. For this
purpose it should be reinforced with all the disposable
light cavalry of the army.

CHANDLER: The role of the advance guard formations in
the attack was very important to Napoleon's strategic
concepts – both those of envelopment and of the central
position. It is noteworthy that he stresses both the reconnaissance and hard fighting tasks required as well as

enlightened leadership at all levels. He also demands a proper balance of forces, as was the case with every French army corps, in order to assure the capability of taking on superior numbers while awaiting the arrival of the rest of the army.

The same principles hold good for an army on the defensive. An active, probing defence of manœuvre by forces in close proximity to the opponent, whether actual or presumed, is far superior to a wholly static one as the French learned to their cost at Dien Bien Phu, and the Israelis holding the Bar Lev line along the eastern bank of the Suez Canal at the outset of the Yom Kippur War (1973), when they were undoubtedly taken by surprise by the sudden Egyptian attack which blasted avenues through the huge sand defences with high-pressure water hoses. Of course, the existence of a fixed frontier – such as the Suez Canal, the 49th parallel in Korea, or the inner German border mine, marsh and wire barrier across Central Europe – poses great problems to active patrolling if charges of provocative actions are to be avoided. The landing of SAS and SBS forces on the Falklands a good month ahead of the arrival of the main expeditionary force provides a good modern example of a small but élite and active advance force carrying out mainly reconnaissance and destabilisation roles to excellent effect.

XXXIII

It is contrary to all the usages of war to allow parks or batteries of artillery to enter a defile, unless you hold the other extremity. In case of retreat, the guns will embarrass your movements and be lost.

NOTHING encumbers the march of an army so much as a quantity of baggage. In the campaign of 1796, Napoleon abandoned his battering-train under the walls of Mantua, after spiking the guns and destroying the carriages. By this sacrifice he acquired a facility of manœuvring rapidly his little army, and obtained the initiative as well as a general superiority over the numerous but divided forces of Marshal Wurmser.

In 1799, during his retreat in Italy, General Moreau being compelled to manœuvre among the mountains, preferred separating himself entirely from his reserve artillery, which he directed upon France by the Col de Fenestrelle, rather than embarrass his march with this part of his equipment.

These are the examples we should follow, for if, by a rapidity of march and a facility of concentration upon decisive points, the victory is gained, the *matériel* of an army is soon re-established. But if, on the other hand, we are beaten and compelled to retreat, it will be difficult to save our equipments, and we may have reason to congratulate ourselves that we abandoned them in time to prevent them from augmenting the trophies of the enemy.

CHANDLER: The dangers of being caught in a cul-de-sac are self-evident. It almost happened to Marshal Marmont at the Festieux Defile to Napoleon's severe displeasure in 1814. In 1800 it was only at great risk that the Army of the Reserve passed a few guns and caissons past Fort Bard amidst the Alpine passes in order to provide Lannes's advance guard (which had bypassed the obstacle) with some artillery, but in that case the

French could claim to hold both extremities of the defile, if not its centre. The appalling fate of the Egyptian armour and transport caught in the Mitla Pass during the Six Day War of 1967 is a convincing modern example.

XXXIV

It should be laid down as a principle, never to leave intervals by which the enemy can penetrate between corps formed in order of battle, unless it be to draw him into a snare.

IN the campaign of 1757 the Prince of Lorraine, who was covering Prague with the Austrian army, perceived the Prussians threatening by a flank movement to turn his right. He immediately ordered a partial change of front by throwing back the infantry of that wing, so as to form a right angle with the rest of the line. But this manœuvre being executed in presence of the enemy, was not effected without some disorder. The heads of the columns having marched too quick, caused the rear to lengthen out, and when the line was formed to the right, a large interval appeared at the salient angle. Frederick observing this error hastened to take advantage of it. He directed his centre corps, commanded by the Duke of Bevern, to throw itself into this opening, and by this manœuvre decided the fate of the battle.

The Prince of Lorraine returned to Prague, beaten and pursued, with the loss of sixteen thousand men, and two hundred pieces of cannon.

It should be observed at the same time that this operation of throwing a corps into the intervals made by an army in line of battle, should never be attempted unless you are at least equal in force, and have an opportunity of outflanking the enemy on the one side or the other; for it is then only you can hope to divide his army in the centre, and isolate the wings entirely. If you are inferior in number, you run the risk of being stopped by the reserves, and overpowered by the enemy's wings, which may deploy upon your flanks and surround you.

It was by this manœuvre that the Duke of Berwick gained the battle of Almanza, in the year 1707, in Spain.

The Anglo-Portuguese army, under the command of Lord Galloway, came to invest Villena. Marshal Berwick, who commanded the French and Spanish army, quitted his camp at Montalegre, and moved upon this town to raise the siege. At his approach, the English general, eager to fight a battle, advanced to meet him in the plains of Almanza. The issue was long doubtful. The first line, commanded by the Duke of Popoli, having been broken, the Chevalier d'Asfeldt, who had charge of the second, drew up his masses with large intervals between them, and when the English, who were in pursuit of the first line, reached these reserves, he took advantage of their disorder to attack them in flank, and defeated them entirely.

Marshal Berwick, perceiving the success of this manœuvre, threw open his front, and deploying upon the enemy's flanks, while the reserve sustained the attack in front, and the cavalry manœuvred in their rear, obtained a complete victory.

Lord Galloway, wounded and pursued, collected, with difficulty, the remains of his army, and took shelter with them in Tortosa.

CHANDLER: The sense of this concept is clear. One only needs to recall the skill of the Japanese at infiltrating forces between British units in Malaya (1941) and Burma (1942–3) to appreciate the dangers involved. Again, the ability of General Student to slip small units of German paratroopers across the single road linking Nijmegen with Arnhem in 1944 gravely hindered the advance of XXX Corps to link up with the British paratroopers holding the bridgehead.

However, the qualification in the maxim's last sentence is significant. To induce the enemy to make a false move leading to his discomfiture – an ancient aim of warfare – there are times when he must be tempted to exploit apparent weaknesses. Modern NATO land doctrine at the operational and tactical levels makes much of luring Soviet attacks into pre-arranged killing-zones formed by valleys whose extremity and sides are strongly held by defending forces. It was just such a blunder that led Feversham's advance guard into a telling ambush in "Bloody Lane" at Philip's Norton in 1685 – or General Dupont into the Spanish trap at Bailen in 1808; indeed, in that case the French commander had aggravated his error by deliberately detaching Vedel's division from the main body. Today, of course, armies generally fight in far more dispersed a fashion than formerly – to avoid air attack or heavy losses from massed artillery fire, and above all so as not to present a tempting target for a tactical nuclear weapon if such a stage has been reached.

XXXV

Encampments of the same army should always be formed so as to protect each other.

AT the battle of Dresden, in the campaign of 1813, the camp of the allies, although advantageously placed upon the heights on the left bank of the Elbe, was nevertheless extremely defective from being traversed longitudinally by a deep ravine which separated the left wing completely from the centre and right. This vicious formation did not escape the penetrating eye of Napoleon. He instantly carried the whole of his cavalry and two corps of infantry against the isolated wing, attacked it with superior numbers; overthrew it, and took ten thousand prisoners before it was possible to come to its support.

CHANDLER: This maxim was disregarded by the Italian Tenth Army at Sidi Barrani in late 1940 and insufficiently applied by the French élite forces holding Dien Bien Phu in Indo-China (1954) – in each case to their severe disadvantage. (See Maxim XXII above.)

XXXVI

When the enemy's army is covered by a river, upon which he holds several têtes de pont, do not attack in front. This would divide your force, and expose you to be turned. . . .

IF you occupy a town or village on the bank of a river opposite to that held by the enemy, it is an advantage to make this spot the crossing point, because it is easier to cover your carriages and reserve artillery as well as to mask the construction of your bridge, in a town, than in the open country. It is also a great advantage to pass a river opposite a village, when this last is only weakly occupied; because as soon as the advanced guard reaches the other side it carries this post, makes a lodgment, and by throwing up a few defensive works, converts it easily into a *tête de pont*. By this means the rest of the army is enabled to effect the passage with facility.

CHANDLER: There are both good examples of, and good exceptions to, this rule. Napoleon's own crossing of the Danube to Lobau Island in May 1809 was covered by a diversion towards Stadlau nearer Vienna. The same ruse was employed again two months later to disguise the French second crossing to the north bank from Lobau immediately prior to Wagram. Similarly, the passage of the Beresina in November 1812 was achieved opposite Studienka (where a ford had been found) while demonstrations towards Ucholodi, several miles to the south, distracted Admiral Tshitsagov's attention. In the Second World War, General Slim fooled the Japanese as to his exact intended crossing places over the Irrawaddy by means of a provocative probe by 19th Division at Kyaukmyaung on 14 January (to which the Japanese responded violently) just before the main crossings elsewhere preceding the great double battle of Mandalay-Meiktila (1945). In Europe two months later, on the Rhine, the surprise capture of an intact bridge at Remagen permitted the Allies to establish a bridgehead to the east of the river several weeks ahead of the major scheduled crossings to the north by Montgomery's 21st Army Group and to the south by Patton's Third Army respectively.

On other occasions rapid movement and surprise have permitted an army to make a safe crossing of a major river obstacle on either a broad or narrow front. Cases of the former are Marlborough's pounce out of a misty dawn on 26 November 1708 upon the four French *têtes de pont* at Gavre, Oudenarde, Kerkhoff and Hauterive on the Scheldt during the operations surrounding the great siege of Lille. The Egyptian crossing of the Suez Canal was also on a broad front on the afternoon of 6 October 1973. Cases of the latter include General Rommel's celebrated passage of the Meuse near Sedan on 14 May 1940, and General Adan's crossing of the Suez Canal with the aid of a large mobile bridge on the night of 17/18 October 1973.

XXXVII

From the moment you are master of a position which commands the opposite bank, facilities are acquired for effecting the passage of the river; above all, if this position is sufficiently extensive, to place upon it artillery in force. . . .

FREDERICK observes that the passage of great rivers in the presence of the enemy is one of the most delicate operations in war. Success on these occasions depends on secrecy, on the rapidity of the manœuvres, and the punctual execution of the orders given for the movements of each division. To pass such an obstacle in presence of an enemy and without his knowledge, it is necessary not only that the previous dispositions should be well conceived, but that they should be executed without confusion. . . . When a river is less than sixty toises (or 120 yards) in breadth, and you have a post upon the other side, the troops which are thrown across derive such advantages from the protection of your artillery that, however small the angle may be, it is impossible for the enemy to prevent the establishment of a bridge. . . .

In the campaign of 1705 Prince Eugene of Savoy, wishing to come to the assistance of the Prince of Piedmont, sought for a favourable point at which to force the passage of the Adda, defended at that time by the French army under the command of the Duke de Vendôme.

After having selected an advantageous situation, Prince Eugene erected a battery of twenty pieces of cannon on a position which commanded the entire of the opposite bank, and covered his infantry by a line of entrenched parallels constructed on the slope of the declivity.

They were working vigorously at the bridge when the Duke de Vendôme appeared with his whole army. At first he seemed determined to oppose its construction, but after having examined the position of Prince Eugene he judged this to be impracticable.

He therefore placed his army out of reach of the prince's batteries, resting both his wings upon the river, so as to form a bow, of which the Adda was the chord. He then covered himself with entrenchments and abbatis, and was thus enabled to charge the enemy's columns whenever they debouched from the bridge and to beat them in detail.

Eugene having reconnoitred the position of the French considered the passage impossible. He therefore withdrew the bridge, and broke up his camp during the night.

It was by this manœuvre also that, in the campaign of 1809, the Archduke Charles compelled the French to reoccupy the Isle of Lobau after having debouched on the left bank of the Danube. The march of the Archduke Charles was wholly concentric. He menaced Gros-aspern with his right, Esling with his centre, and Enzersdorf with his left.

His army, with both wings resting on the Danube, formed a semicircle round Esling. Napoleon immediately attacked and broke the centre of the Austrians, but after having forced their first line he found himself arrested by the reserves. In the meantime, the bridges upon the Danube had been destroyed, and several of his corps, with their parks of artillery, were still on the right bank. This disappointment, joined to the favourable position of the Austrians, decided Napoleon to re-enter the Isle of Lobau, where he had previously constructed a line of field-works, so as to give it all the advantages of a well-entrenched camp.

CHANDLER: Napoleon had much personal experience of the problems of crossing bridges and rivers in the proximity of the enemy. In 1796 he faced such problems at Lodi and Arcola; in 1806 at Jena; in 1809 at Aspern-Essling and Wagram; in 1812 at the Beresina; next year at Leipzig; and in 1815 at Charleroi. The need to seize a small bridgehead on the further bank prior to enlarging its size remains as true today for a Soviet armoured or motor-rifle division – or a NATO force – as ever it did some 200 years ago when Napoleon seized the Isle of Lobau adjacent to the north bank of the Danube – even if the initial assault is likely to be launched by heliborne stormtroops rather than by bold rushes or small boat-parties of breathless grenadiers and darting light infantrymen. The need for heavy covering fire – today provided by air strikes or massed long-range artillery fire rather than by short-ranged cannon – is still axiomatic, except in such rare cases as Murat's surprise of the bridge at Vienna in 1805 or Engeman's and Timmer-mann's similar feat at Remagen 140 years later. The actions of Archduke Charles in attempting to seal off the French Danubian bridgehead in the hope of defeating the enemy in detail as they strove to pass the river led to a major repulse for Napoleon at Aspern-Essling in 1809, although his second attempt was ultimately successful at Wagram.

One of the geophysical problems that has always faced would-be invaders of Russia from the west has been the tendency for eastern banks of the great Russian rivers to overlook the western banks.

XXXVIII

It is difficult to prevent an enemy, supplied with pontoons, from crossing a river. When the object of an army which defends a river is to cover a siege, the moment the general has ascertained his inability to oppose the passage, he should take measures to arrive before the enemy at an intermediate position between the river he defends and the place he desires to cover.

HERE we may observe that this intermediate position should be reconnoitred, or rather, well entrenched beforehand; for the enemy will be unable to make an offensive movement against the corps employed in the siege until he has beaten the army of observation; and this last, under cover of its camp, may always await a favourable opportunity to attack him in flank or in rear.

Besides, the army which is once entrenched in this manner has the advantage of being concentrated; while that of the enemy must act in detachments, if he wishes to cover his bridge, and watch the movements of the army of observation so as to enable him to attack the besieging corps in its lines, without being exposed to an attempt on his rear, or being menaced with the loss of his bridge.

CHANDLER: Many classical and modern battles have been fought in association with the conduct – and attempted reliefs – of major sieges, and rivers have often been involved in these instances. The duty of the covering force to interpose itself between the relief army and the siege lines has generally been clear: Marlborough's covering of Prince Eugene's siege of Lille (1708) is one good example, showing many variations on the theme. Napoleon's conduct of the siege of Mantua (1796–7), twice broken-off by the need to head off Austrian relief armies under Generals Würmser and d'Alvintzi, is another. Once the Austrians had passed over the Adige

near Verona in January 1797, Napoleon had to strain
every nerve – despite having defeated d'Alvintzi at Rivoli
– to interpose his hard-marching army at La Favorita to
prevent reinforcements and supplies being passed in to
Würmser's garrison. This he achieved, and Würmser at
last surrendered.

XXXIX

*In the campaign of 1645, Turenne was attacked with his army
before Philipsburg by a very superior force. There was no bridge
here over the Rhine, but he took advantage of the ground between
the river and the place to establish his camp. . . .*

MARSHAL Saxe, in the campaign of 1741, having passed
the Moldau in quest of a detached corps of fourteen
thousand men, which was about to throw itself into
Prague, left a thousand infantry upon that river, with
orders to entrench themselves upon a height directly
opposite the *tête de pont*.

By this precaution the marshal secured his retreat, and
also the facility of repassing the bridge without disorder,
by rallying his divisions between the entrenched height
and the *tête de pont*.

Were these examples unknown to the generals of
modern times, or are they disposed to think such precautions superfluous?

CHANDLER: Here one thinks of Feversham's leaving a
sufficient interval between his camp and the Bussex
rhyne at Westonzoyland in July 1685 before the night
attack that led to the battle of Sedgemoor. The great

engineer, Vauban, designed the defences of Toulon to incorporate a line of detached forts at a distance from the main fortress so as to permit a defending army, if present within the siege lines, to play an active part in the defence. Prince Eugene, besieging Belgrade in 1717, faced the same problem as Cæsar at Alesia (52BC), namely finding himself hemmed in between the garrison within the fortress and a large relieving army without. Both commanders, as befitted their inclusion in Napoleon's select list of the Seven Great Captains, overcame the problem triumphantly by leaving themselves sufficient manœuvring room between their different opponents. Armies retiring over rivers obviously need room between the defences of the *tête de pont* and their bridges, as the Turks learned to their cost at Zenta on 11 September 1697, where Mustafa II lost 20,000 killed and 10,000 drowned. On a larger scale – with the English Channel at his back and evacuation his only hope, Lord Gort depended upon the Dunkirk perimeter's being large enough to hold off the Germans for the necessary time between 28 May and 4 June 1940.

XL

Fortresses are equally useful in offensive as in defensive warfare. It is true they will not in themselves arrest an army, but they are an excellent means of retarding, embarrassing, weakening and annoying a victorious enemy.

THE brilliant success of the allied armies in the campaign of 1814 has given to many military men a false idea of the real value of fortresses.

The formidable bodies which crossed the Rhine and the Alps at this period were enabled to spare large detachments to blockade the strong places that covered the frontiers of France, without materially affecting the numerical superiority of the army which marched upon the capital. This army was in a condition, therefore, to act, without the fear of being menaced in its line of retreat.

But at no period of military history were the armies of Europe so combined before, or governed so entirely by one common mind in the attainment of a single object. Under these circumstances, the line of fortresses which surround France was rendered unavailable during the campaign; but it would be very imprudent, therefore, to conclude that a frontier guarded by numerous fortresses may be passed with impunity, or that battles may be fought with these places in your rear without previously besieging, or at least investing them with sufficient forces.

CHANDLER: Perhaps the classic example of this precept is the Quadrilateral in North Italy – the four fortresses of Mantua and Peschiera on the River Mincio, and of Legnagno and Verona on the River Adige further east. These positions guarded the exits from the Alpine passes to the north and east and also protected the approaches to Lake Garda and the River Po. They all figured large in the history of the campaign of 1796. Clearly, to be fully effective in harrassing a would-be attacker, a field force operating between and around them is also necessary.

Fortresses continue to play an active part in modern warfare, both in offence and (above all) in defence. "Fortress Stalingrad" was the lynch-pin of the Red Army's defence of the Volga in 1942, and the spring-board for their great counter-offensive launched in early 1943 by acting as a fatal magnet to von Paulus's doomed Sixth Army. Similarly, in the wars in the Western Desert, 1940–3, the seaport-fortress of Tobruk played a vital part in both German and British plans. The role played by Monte Cassino on its massif in the Italian Campaign was also for many months of 1944 a deter-minant factor, as was the huge and terrible siege of Leningrad on the Eastern Front, 1941–3.

In any future major European struggle, NATO would rely on holding the great connurbations of West Germany fatally to ensnare Soviet tank armies, while tiny village strongholds would be held by small last-ditch garrisons far to the rear so as to harass the enemy's communications and hence his resupply and reinforce-ment capabilities.

XLI

There are only two ways of ensuring the success of a siege. The first, to begin by beating the enemy's army employed to cover the place, forcing it out of the field, and throwing its remains beyond some great natural obstacle, such as a chain of mountains or large river. . . .

WHEN we undertake a siege, says Montecuccoli, we should not seek to place ourselves opposite the weakest part of the fortress, but at the point most favourable for establishing a camp and executing the designs we have in view. This maxim was well understood by the Duke of Berwick.

Sent to form the siege of Nice in 1706, he determined to attack on the side of Montalban, contrary to the advice of Vauban, and even to the orders of the king. Having a very small army at his disposal, he began by securing his camp. This he did by constructing redoubts upon the heights that shut in the space between the Var and the Paillon, two rivers which supported his flanks. By this means he protected himself against a surprise, for the Duke of Savoy having the power of debouching suddenly by the Col de Tende, it was necessary that the marshal should be enabled to assemble his forces so as to move rapidly upon his adversary and fight him before he got into position, otherwise his inferiority in numbers would have obliged him to raise the siege.

When Marshal Saxe was besieging Brussels with only twenty-eight thousand men opposed to a garrison of twelve thousand, he received intelligence that the Prince of Waldeck was assembling his forces to raise the siege. Not being strong enough to form an army of observation, the marshal reconnoitred a field of battle on the little river Voluve, and made all the necessary dispositions for moving rapidly to the spot in case of the approach of the enemy. By this means he was prepared to receive his adversary without discontinuing the operations of the siege.

CHANDLER: This sequence of maxims devoted to siege warfare is interesting but at the same time rather suspicious. Napoleon had little time for the delays associated with siege warfare – at least, after Mantua in 1796–7 when he learnt at first hand the disadvantages of surrendering the initiative to the enemy by becoming tied to a major siege in progress with only limited resources available. These maxims, then, could possibly be the inspiration of "another compiler" – perhaps even Balzac (see page 32). They embody the best 17th and 18th century practice but are rather un-Napoleonic.

The interaction between besieging force and covering army is well described. Marlborough's great sieges of Lille (1708) and Bouchain (1711) are excellent examples of the second type listed, although it was French, not Allied, hesitations that prevented a great battle on each occasion. Nevertheless, General Webb's spirited action at Wynendael in 1708 was closely linked to events before Lille. At Mantua, Napoleon time and again had to intervene to head off would-be relief forces, just as Marshal Beresford had to fight at Albuera in 1811 to protect the siege of Badajoz from Soult's relieving army, and Wellington at Fuentes de Oñoro to safeguard that of Almeida from Massena. General Grant's famous siege of Vicksburg (1863) is an excellent example of a commander abandoning his communications by keeping adequate supplies with his army. Prince Eugene's conduct of the siege of Belgrade (1717) is a fine instance of a commander faced by very long odds keeping his besieging and covering forces united against the Turkish garrison and their relieving army.

Relief columns (rather than whole armies) are more often involved in modern times: Auchinleck's success in relieving the second siege of Tobruk (1941); Manstein's failure to break through to Stalingrad in 1942/3; Patton's success at relieving Bastogne on Christmas Day, 1944; and the failure of the French column to reach Dien Bien Phu in 1954. The most dramatic "relief" of modern times was undoubtedly that of Leningrad, which was rescued from the east by the Red Army laying a railway line and creating a road over the frozen surface of Lake Ladoga (1942/3) to enable reinforcements and supplies to reach the starving population and garrison. Air supply also failed to save von Paulus in Stalingrad (despite Goering's flamboyant promises), but it certainly helped win the double-siege/battle of Kohima-Imphal in Burma (1944) and aided the Arnhem garrison to hold out for longer than would otherwise have been the case in North-West Europe in the September of that year.

However, there are few examples of major battles being fought in direct relief of a modern siege – except possibly Operation "Crusader" in the Western Desert in late 1941, linked with the siege of Tobruk.

XLII

Feuquières says that we should never wait for the enemy in the lines of circumvallation, but that we should go out and attack him. He is in error. . . .

DURING the siege of Mons in 1691, the Prince of Orange assembled his army and advanced as far as Notre Dame de Halle, making a demonstration to succour the place. Louis XIV, who commanded the siege in person, called a council of war to deliberate on what was to be done in case the Prince of Orange approached. The opinion of

Marshal Luxembourg was to remain within the lines of circumvallation, and that opinion prevailed.

The marshal laid it down as a principle, that when the besieging army is not strong enough to defend the whole extent of circumvallation, it should quit the lines and advance to meet the enemy; but when it is strong enough to encamp in two lines around a place, that it is better to profit by a good entrenchment, more especially as by this means the siege is not interrupted.

In 1658 Marshal Turenne was besieging Dunkirk. He had already opened the trenches when the Spanish army, under the orders of the Prince Don Juan, Condé, and d'Hocquincourt, appeared in sight, and took post upon the Downs at the distance of a league from his lines. Turenne had the superiority in numbers, and he determined to quit his entrenchments. He had other advantages also. The enemy were without artillery, and their superiority in cavalry was rendered useless by the unfavourable nature of the ground. It was therefore of great importance to beat the Spanish army before it had time to entrench itself and bring up its artillery. The victory gained by the French on this occasion justified all the combinations of Marshal Turenne.

When Marshal Berwick was laying siege to Philipsburgh in 1733, he had reason to apprehend that the Prince of Savoy would attack him with all the forces of the empire before its termination. The marshal therefore, after having made his disposition of the troops intended for the siege, formed with the rest of his army a corps of observation to make head against Prince Eugene, in case the latter should choose to attack him in his lines, or attempt a diversion on the Moselle or Upper Rhine. Prince Eugene having arrived in front of the besieging army, some general officers were of opinion that it was better not to await the enemy within the lines, but to move forward and attack him. But Marshal Berwick, who agreed with the Duke of Luxembourg, that an army which can occupy completely good entrenchments is not liable to be forced, persisted in remaining within his works.

The result proved that this was also the opinion of Prince Eugene, for he did not dare to attack the entrenchments, which he would not have failed to do if he had had any hopes of success.

CHANDLER: Sorties in strength have their place in sieges, ancient and modern, but there should indeed be no hard and fast rule applied. However, Napoleon usually erred on the side of opportunism and action – and it is somewhat surprising to find him advocating caution in this instance. All great sieges, such as those of Paris (1870–1), Plevna (1877) and Port Arthur (1904), saw considerable use of sorties to gain time, sap the besiegers' morale and boost that of the besieged, just as large-scale efforts were made by von Paulus at Stalingrad and Scoones at Imphal. Only at Dien Bien Phu did the garrison show little taste for true offensive action. Vauban, doyen of siege warfare, was adamant about the need for the besieged to sally forth on occasion and also for the besiegers to take strong measures against would-be relief armies. As always, therefore, everything depends on the precise circumstances: at Stalingrad the Russians pre-

vented Manstein's break-in from the South; at Dien Bien
Phu, Giap was equally successful against the French
relief attempt.

XLIII

Those who proscribe lines of circumvallation, and all the assis-
tance which the science of the engineer can afford, deprive them-
selves gratuitously of an auxiliary which is never injurious, almost
always useful, and often indispensable. . . .

IF we are inferior in numbers, says Marshal Saxe,
entrenchments are of no use, for the enemy will bring all
his forces to bear upon particular points. If we are of
equal strength, they are unnecessary also. If we are
superior we do not want them. Then why give ourselves
the trouble to entrench? Notwithstanding this opinion of
the inutility of entrenchments, Marshal Saxe had often
recourse to them.

   In 1797, Generals Provéra and Hohenzollern having
presented themselves before Mantua (where Marshal
Würmser was shut up) for the purpose of raising the
siege, they were stopped by the lines of contravallation of
St. George. This slight obstacle sufficed to afford
Napoleon time to arrive from Rivoli and defeat their
enterprise. It was in consequence of neglecting to
entrench themselves that the French had been obliged to
raise the siege in the preceding campaign.

CHANDLER: Lines of circumvallation (facing outwards
from the place being besieged) are clearly of value for
achieving economy of force, and guarding against the

possibility of surprise relief parties breaking through to the succour of the besieged garrison. Napoleon's strictures about the poor state of field engineering led to considerable notice being taken of the subject by Germans and others in the 19th century, German *feste* redoubts at Metz being one example and their generally superior trenches and dug-outs in 1914–18, another. It took the Great Boer War of 1899–1902 to inspire equivalent interest in the British Army, particularly the strong Boer positions defending Magersfontein, but considerable leeway had been made up by 1914, the first *Manual of Field Engineering* being published in about 1905. Today all armies take great care to develop and modernise the science and art of field engineering, although yet again it was deficiencies in the design of defences at Dien Bien Phu (particularly in respect of strong enough overhead cover) that materially prepared the way for the French disaster in 1954. Argentinian prepared positions in the Falklands War were considered to have been well constructed.

XLIV

If circumstances prevent a sufficient garrison being left to defend a fortified town which contains a hospital and magazines, at least every means should be employed to secure the citadel against a coup de main.

A FEW battalions dispersed about a town inspire no terror, but shut up in the more narrow outline of a citadel, they assume an imposing attitude. For this reason it appears to me that such a precaution is always necessary, not only in fortresses, but wherever there are hospitals, or depôts of any kind. Where there is no citadel, some quarter of the town should be fixed upon most favourable for defence, and intrenched in such a manner as to oppose the greatest resistance possible.

CHANDLER: Again the expression of the patently obvious – but then many truths bear repetition. Examples include the French defence of the three forts in Salamanca (1812) against Wellington (who had to tie down a division for valuable days as a result), and the Japanese defence of Fort Dufferin at Mandalay against Slim's Fourteenth Army (1945); and above all the North Vietnamese and Viet-Cong garrisoning of the ancient citadel at Hué in South Vietnam during the Tet Offensive of 1968. It took a US Marine Corps a month to regain control of the city and fortress and cost many lives and much destruction to the city in the process.

XLV

A fortified place can only protect the garrison and arrest the enemy for a certain time. When this time has elapsed, and the defences are destroyed, the garrison should lay down its arms. All civilised nations are agreed on this point, and there has never been an argument except with reference to the greater or less degree of defence which a governor is bound to make before he capitulates. . . .

IN 1705 the French, who were besieged in Haguenau by Count Thungen, found themselves incapable of sustaining an assault. Péri, the governor, who had already distinguished himself by a vigorous defence, despairing of being allowed to capitulate on any terms short of becoming prisoners of war, resolved to abandon the place, and cut his way through the besiegers.

In order to conceal his intention more effectually, and while he deceived the enemy to sound at the same time the disposition of his officers, he assembled a council of war, and declared his resolution to die in the breach. Then, under pretext of the extremity to which he was reduced, he commanded the whole garrison under arms, and leaving only a few sharpshooters in the breach, gave the order to march, and set out in silence under cover of the night from Haguenau. This audacious enterprise was crowned with success, and Péri reached Saverne without having suffered the smallest loss.

Two fine instances of defence in later times are those of Massena at Genoa and of Palafox at Saragossa.

The first marched out with arms and baggage and all the honours of war, after rejecting every summons, and defending himself until hunger alone compelled him to capitulate. The second only yielded after having buried his garrison amid the ruins of the city, which he defended from house to house until famine and death left him no alternative but to surrender. This siege, which was equally honourable to the French as to the Spaniards, is one of the most memorable in the history of war. In the course of it Palafox displayed every possible resource which courage and obstinacy can supply in the defence of a fortress.

All real strength is founded in the mind; and on this account I am of opinion that we should be directed in the choice of a governor less by his genius than his personal character. His most essential qualities should be courage, perseverance, and soldier-like devotedness. Above all, he should possess the talent, not only of infusing courage into the garrison, but of kindling a spirit of resistance in the whole population. Where this last is wanting, however art may multiply the defences of a place, the garrison will be compelled to capitulate after having sustained the first or at most the second assault.

CHANDLER: This maxim is very un-Napoleonic. He was far less impressed with the "civilised" conventions of siege warfare than his predecessors. Thus General Phillipon held Badajoz against Wellington to the bitter end in 1812, requiring a full-scale storming to be carried out with heavy loss and horrible after-effects for the Spanish population once the city was taken. The same was true of the two sieges of Burgos and that of San Sebastien (1813). A model example of a garrison successfully evacuating a hopelessly compromised fortress was that of General Brennier's escape with 1,300 men from besieged Almeida on the night of 10 May 1811 – an event that earned Wellington's gravest ire against the Allied commanders held responsible. In the Second World War, the successful evacuation of the "boxes" held by XIII Corps at Gazala and the escape of General Koenig from Bir Hacheim (1942) slightly redeemed the Allied failure in a series of operations that earned Rommel his field marshal's baton. The South African garrison of Gazala itself, however, was forced to surrender, and many supplies and much petrol thus fell into German hands.

XLVI

The keys of a fortress are always well worth the retirement of the garrison when it is resolved to yield only on those conditions. On this principle it is always wiser to grant an honourable capitulation to a garrison which has made a vigorous resistance then to risk an assault.

MARSHAL Villars has justly observed that no governor of a place should be permitted to excuse himself for surrendering, on the ground of wishing to preserve the king's troops. Every garrison that displays courage will escape being prisoners of war. For there is no general who, however well assured of carrying a place by assault, will not prefer granting terms of capitulation rather than risk the loss of a thousand men in forcing determined troops to surrender.

CHANDLER: Again, this is much more of an 18th century concept than a 19th or 20th century one. The French did not grant generous terms to the garrisons of Mantua (1796–7), Danzig (1807) or Saragossa (1808–9); nor did their garrisons request or accept them at Ciudad Rodrigo, Badajoz (both 1812) or San Sebastien (1813), at least not until their respective towns had been put to the storm. Thereafter both the Badajoz citadel on the North bank of the Guadiana and the citadel atop Mount Orgueil at San Sebastien did capitulate on the best terms available. That also happened at Burgos in 1813. But capitulations on terms were much more part of the

Marlburian scene in the early 18th century: Marshal Boufflers was allowed to march out and retire from Lille's citadel in 1708 after sustaining a herculean siege. There are fewer examples of garrisons "marching away to fight another day" in more recent times. Napoleon granted terms to the Mamelukes holding El Arish Fort in 1799 – only to massacre them in cold blood when he found the same soldiers taking part in the defence of Jaffa a few weeks later. In the First World War the gallant German commander in East Africa, General Lettow von Vorbeck, was permitted to march out to surrender with bayonets fixed and colours flying in late 1918 as a mark of Allied respect for his superb fighting qualities against overwhelming odds. Most other capitulations, as at Stalingrad, Dien Bien Phu or Port Stanley (1982), have only granted the survivors discretion to lay down their arms as prisoners of war – in other words to accept the victors' terms without question.

XLVII

Infantry, cavalry, and artillery are nothing without each other. They should always be so disposed in cantonments as to assist each other in case of surprise.

A GENERAL, says Frederick, should direct his whole attention to the tranquillity of his cantonments, in order that the soldier may be relieved from all anxiety, and repose in security from his fatigues. With this view, care should be taken that the troops are able to form rapidly upon ground which has been previously reconnoitred; that the generals remain always with their divisions or brigades, and that the service is carried on throughout with exactness.

Marshal Saxe is of opinion that an army should not be in a hurry to quit its cantonments, but that it should wait till the enemy has exhausted himself with marching, and be ready to fall upon him with fresh troops when he is overcome with fatigue.

I believe, however, that it would be dangerous to trust implicitly to this high authority, for there are many occasions where all the advantage lies in the initiative, more especially when the enemy has been compelled to extend his cantonments from scarcity of subsistence, and can be attacked before he has time to concentrate his forces.

CHANDLER: The requirements of carefully combined all-arms forces have indeed become a tenet of all warfare since the 18th century. Mutual support both in defence and attack is a cardinal principle. Examples when disasters have occurred through neglect of this concept range from Ney's all-cavalry onslaughts against the Allied squares at Waterloo (1815), to the commander of the 51st (Highland) Division's refusal to adopt Colonel Fuller's prescribed tank-infantry tactical drills at Cambrai (1917), or the fate of the Israeli armoured brigade that counter-attacked on 8 October 1973 without support shortly after the opening of the Yom Kippur War and lost two-thirds of its tanks in some thirty minutes of action, mainly to small anti-tank missiles handled by Egyptian infantry with impunity in the absence of Israeli infantry to fight them off. All-arm co-operation, to include an integrated use of airpower and helicopters, is a vital consideration in all modern operations, as was seen in the Falklands' land campaign.

XLVIII

The formation of infantry in line should be always in two ranks, because the length of the musket only admits of an effective fire in this formation. The discharge of the third rank is not only uncertain, but frequently dangerous to the ranks in its front. . . .

I AM of opinion, if circumstances require a line of infantry to resort to a square, that two deep is too light a formation to resist the shock of cavalry. However useless the third rank may appear for the purpose of file-firing , it is, notwithstanding, necessary, in order to replace the men who fall in the ranks in front; otherwise you would be obliged to close in the files, and by this means leave intervals between the companies which the cavalry would not fail to penetrate. It appears to me, also, that when infantry is formed in two ranks the columns will be found to open out in marching to a flank. If it should be considered advantageous behind entrenchments to keep the infantry in two ranks, the third rank should be placed in reserve and brought forward to relieve the front rank when fatigued, or when the fire is observed to slacken. I am induced to make these remarks, because I have seen an excellent pamphlet which proposes the two-deep formation for infantry as the best. The author supports his opinion by a variety of plausible reasons, but not sufficient, as it appears to me, to answer all the objections that may be offered to this practice.

CHANDLER: This is a case of Napoleon learning wisdom "after the event". His infantry almost always fought three lines deep; it was Wellington who insisted on only the double line in the Peninsula and at Waterloo, following the tenets laid down by General Dundas in the drill book (*c*.1798) used by the British Army of the period. Except in so far as it has an "economy of force" implication and teaches the eternal lesson of the need to keep a reserve to watch the flanks and rear, this maxim has clearly little significance today. The switch from muzzle-loading smooth-bore muskets to the rifled breach-loaders of the middle to late 19th century and the advent of smokeless powder and magazine-loading weapons, and above all of machine-guns and barbed wire, revolutionised infantry tactics.

XLIX

The habit of mixing small bodies of infantry and cavalry together is a bad one, and attended with many inconveniences. The cavalry loses its power of action. It becomes fettered in all its movements. Its energy is paralysed. . . .

THIS also was the opinion of Marshal Saxe. The weakness of the above formation, says he, is sufficient in itself to intimidate the platoons of infantry, because they must be lost if the cavalry is beaten.

The cavalry, also, which depends on the infantry for succour, is disconcerted the moment a brisk forward movement carries them out of sight of their supports. Marshal Turenne, and the generals of his time, sometimes employed this order of formation; but that does not, in my opinion, justify a modern author for recommending it in an essay entitled, *Considérations sur l'Art de la Guerre*. In fact, this formation has long been abandoned, and since the introduction of light artillery it appears to me almost ridiculous to propose it.

CHANDLER: All depends here on the definition of "small". The admixture of infantry and cavalry in correct quantities was an important aspect of the grand tactics employed by Gustavus Adolphus, Marlborough and Fredrick the Great – not to forget Napoleon himself. Clearly "penny packets" have little to recommend them. One reason for the cataclysmic defeat of France in 1940 was the failure to keep a strategic reserve in being but instead to use the tanks available in small detachments to bolster the collapsing front. What could be achieved by a properly balanced force was well demonstrated at Arras on 21 May when two "infantry-tank" battalions, as many more infantry battalions and two batteries of Royal Artillery made a telling (if short-lived) counterattack against the flank of the German advance, causing even Rommel much anxiety.

Today the same rules apply, except possibly in the concept of "village strongpoint" defence on the Central Front, where small, mixed garrisons of infantry tanks and guns are intended to cause disproportionate problems to Warsaw Pact forces and their lines of supply. So, yet again, "circumstances alter cases".

L

Charges of cavalry are equally useful at the beginning, the middle, and the end of a battle. They should be made always, if possible, on the flanks of the infantry, especially when this last is engaged in front.

THE Archduke Charles, in speaking of cavalry, recommends that it should be brought in mass upon a decisive point when the moment for employing it arrives; that is to say, when it can attack with a certainty of success. As the rapidity of its movement enables cavalry to act along the whole line in the same day, the general who commands it should keep it together as much as possible, and avoid dividing it into many detachments. When the nature of the ground admits of cavalry being employed on all points of the line, it is desirable to form it in column behind the infantry, and in a position from whence it may be easily directed wherever it is required. If cavalry is intended to cover a position, it should be placed sufficiently in the rear to meet at full speed any advance of troops coming to attack that position. If it is destined to cover the flank of the infantry, it should for the same reason be placed directly behind it. As the object of cavalry is purely offensive, it should be a rule to form it at such a distance only from the point of collision as to enable it to acquire its utmost impulse and arrive at the top of its speed into action. With respect to the cavalry reserve this should only be employed at the end of a battle, either to render the success more decisive or to cover the retreat. Napoleon remarks that at the battle of Waterloo the cavalry of the Guard, which composed the reserve, was engaged against his orders. He complains of having been deprived from five o'clock of the use of this reserve, which, when well employed, had so often ensured him the victory.

CHANDLER: This famous Napoleonic maxim highlights the need to fight fluidly at all stages of a war, campaign or engagement. The use of "shock action" by cavalry, whether heavy cuirassiers, intermediate dragoons, or light hussars and lancers, in co-ordination with infantry and guns was a fundamental condition of Napoleonic tactics and indeed of most of his opponents. Today the same principle applies whether involving main battle tanks, medium tanks and Scorpions, or armed APCs. And, of course, their role in supporting infantry is as important as ever. "Armour conquers, infantry occupies" is a 1939–45 adage that still has much validity. But today the air and heliborne elements have always to be taken into account in any serious engagement.

LI

It is the business of the cavalry to follow up the victory, and to prevent the enemy from rallying.

VICTOR or vanquished, it is of the greatest importance to have a body of cavalry in reserve, either to take advantage of victory or to secure a retreat. The most decisive battles lose half their value to the conqueror when the want of cavalry prevents him from following up his success and depriving the enemy of the power of rallying.

When a retiring army is pursued, it is more especially upon the flanks that the weight of cavalry should fall if you are strong enough in that arm to cut off his retreat.

CHANDLER: In the days of horsed cavalry the pursuit was particularly the province of the dragoons and light cavalry, the intention being to capitalise on success and to allow the beaten foe no opportunity to rally and reorganise his demoralised forces. Marlborough leading the pursuit in person for 15 miles after Ramillies (1706) and completing a 15-hour day in the saddle in his late 50s; Kellermann's dragoons pursuing Moore's army to Corunna; Marshal Murat, riding-whip in hand, leading the French cavalry into Weimar after the battle of Jena; and the Prussian cavalry taking over the pursuit from Wellington's exhausted Allied army after Waterloo; all are examples that make the point. In 1918 it was the Commonwealth Cavalry Corps that mainly exploited General Allenby's great victory over Liman von Sanders's three Turkish armies at Meggido, covering up to 40 miles a day. It was also the province of cavalry to cover a retiring or retreating army; the North British Dragoons (the Scots Greys) forming the rearguard on 17 June 1815 as the Allies fell back from Quatre Bras to Mont St Jean is a classical case.

Today the pursuit role is being increasingly taken over by aircraft and helicopter gunships, as the Soviets have demonstrated in their more successful operations in Afghanistan. But light armour still retains important ground roles in both pursuit and retreat. In 1940 Rommel's "Ghost" Seventh Panzer Division exploited the collapse of the French Ninth Army by pushing boldly ahead (with a temporary setback at Arras on 21 May) all the way to the mouth of the River Somme. Israeli armour also complemented the air exploitation of their successes in Sinai and on the Golan during the Six Day War of 1967.

LII

Artillery is more essential to cavalry than to infantry, because cavalry has no fire for its defence, but depends on the sabre. . . .

HORSE artillery is an invention of Frederick. Austria lost no time in introducing it into her armies, although in an imperfect degree. It was only in 1792 that this arm was adopted in France, where it was brought rapidly to its present perfection.

The services of this arm during the wars of the Revolution were immense. It may be said to have changed to a certain extent the character of tactics, because its facility of movement enables it to bear with rapidity on every point where artillery can be employed with success. Napoleon has remarked in his memoirs that a flanking battery which strikes and rakes the enemy obliquely is capable of deciding a victory in itself. To this we may add, that independent of the advantages which cavalry derives from horse artillery in securing its flanks, and in opening the way for a successful charge by the destructiveness of its fire, it is desirable that these two arms should never be separated, but ready at all times to seize upon points where it may be necessary to employ cannon. On these occasions the cavalry masks the march of the artillery, protects its establishment in position, and covers it from the attack of the enemy until it is ready to open its fire.

CHANDLER: Although cavalry may be said to have had some firepower capacity (pistols and carbines) and dragoons were, in their dismounted role, armed with short-barrelled muskets, there is some truth in this maxim. Prince Eugene of Savoy and Frederick the Great both created horse artillery units to make good the fire imbalance, and the development of "flying artillery" (whose gunners all rode alongside the enlarged gun teams instead of clinging to the limbers) was a further important development. The King's Troop, Royal Horse Artillery, is equipped in this fashion for wholly ceremonial purposes. Such guns (often six-pounders) were invaluable in both pursuit or retreat. Unsupported cavalry were particularly vulnerable while rallying after a charge – as the counter-attack by cuirassiers and lancers against the (unsupported) Scots Greys and the rest of the Union Brigade at 2.30 p.m. at Waterloo demonstrated all too vividly. The previous day, the R.H.A. (including Captain Mercer) had shared the final rearguard duty with the Scots Greys, unlimbering and relimbering, firing and retiring, ridge by ridge, before the pursuing French.

Today, in the age of armour, tanks carry both main and secondary weaponry. However, the restricted vision of tank crews makes them vulnerable to anti-tank weapons and guided missiles, and so both require artillery and infantry support, the former often taking the form of self-propelled guns (true descendants of horse artillery), the latter being conveyed in half-tracks (e.g., Rommel's Panzer-Grenadiers) or armoured personnel carriers, APCs). The present trend is for almost all divisions to become mechanised or "motor-rifle", and mobile artillery makes up a most important component of both.

LIII

In march or in position the greater part of the artillery should be with the divisions of infantry and cavalry. The rest should be in reserve. . . .

THE better infantry is the more important it is to support it by artillery, with a view to its preservation.

It is essential also that the batteries attached to divisions should march in the front, because this has a strong influence on the *moral* of the soldier. He attacks always with confidence when he sees the flanks of the column well covered with cannon.

The artillery reserve should be kept for a decisive moment, and then employed in full force, for it will be difficult for the enemy at such a time to presume to attack it.

There is scarcely an instance of a battery of sixty pieces of cannon having been carried by a charge of infantry or cavalry, unless where it was entirely without support, or in a position to be easily turned.

CHANDLER: In the 17th and much of the 18th century it was the custom for the "train of artillery" to move on campaign as an entity, deploying its guns to specific formations when the line of battle was formed. An exception to this was British, Dutch and Swedish practice of permanently attaching light guns to brigades and regiments (the historical equivalent to the modern infantry support companies). Napoleon did much to fuse advancing, fighting and pursuing into one continuous, remorseless process, and hence it became customary to attach the main artillery batteries to individual corps and divisions from the outset, as remains the modern practice. However, the artillery reserve of 12-pounders remained a force apart, kept under army command for special roles, just as today nuclear artillery in NATO is strictly controlled by SACEUR.

The keeping of sufficient ammunition in ready-use chests, caissons, and wagons close by the pieces was obviously important. The artillery of the French Imperial Guard had a double allocation per cannon. This all represented an important logistic problem. Today the provision of palletted artillery ammunition on active service remains a major problem even in the age of air resupply. In the Falklands the loss of all but one Chinook helicopter on *Atlantic Conveyor* almost compromised the successful conclusion of the land campaign. Major-General Moore's gunners were down to their last few rounds with little prospect of rapid resupply when the Argentinians surrendered. The deficiencies of ammunition resupply have been brought out time and again on the NATO Central Front major autumn exercises. This is a cause for grave concern, as Warsaw Pact and Soviet doctrine place great emphasis on opening an offensive with a daunting hurricane of some 72 hours' fire by massed guns on a truly daunting scale, spearheaded by air-to-ground onslaughts and preceded by *Spetznatz* (Special Forces) attacks in depth against headquarters and supply dumps. The need for a large allocation of ammunition available for immediate use is therefore of critical importance, calling for the bringing in and advance dumping of large quantities. The importance of not losing these resources in any early "take-the-strain"

NATO withdrawal reinforces the arguments of "hawks" who would prefer to see a more aggressive Central Front strategy of the "Air-Land Battle" category.

LIV

Artillery should always be placed in the most advantageous positions, and as far in front of the line of cavalry and infantry, without compromising the safety of the guns, as possible. . . .

THE battery of eighteen pieces of cannon which covered the centre of the Russian army at the battle of La Moskwa (Borodino) may be cited as an example.

Its position upon a circular height which commanded the field in every direction, added so powerfully to its effect that its fire alone sufficed for a considerable time to paralyse the vigorous attack made by the French with their right. Although twice broken, the left of the Russian army closed to this battery, as to a pivot, and twice recovered its former position. After repeated attacks, conducted with a rare intrepidity, the battery was at length carried by the French, but not till they had lost the *élite* of their army, and with it the generals Caulincourt and Montbrun. Its capture decided the retreat of the Russian left.

I might advert likewise to another instance in the campaign of 1809, and to the terrible effect produced by the hundred pieces of cannon of the guard which General Lauriston directed at the battle of Wagram against the right of the Austrian army.

CHANDLER: Guns were bulky items even in historical times, difficult to deploy and comparatively short-ranged, firing over open sights to ranges of up to a mile. Hence this maxim. Napoleon made great use of massed batteries at battles such as Jena, Wagram and Waterloo. However, since the introduction in the mid-19th century of rifled cannon with far longer ranges and the consequent introduction by 1900 of indirect fire, the position of the guns has tended to be behind the front-line formations or well to the rear (in the case of strategic long-range artillery 15 miles and more). However, the modern NATO need is for anti-tank guided weapons (ATGWs) to be sited well forward in profusion to compensate for the Warsaw Pact's three-to-one superiority in main battle tanks (MBTs). The Egyptian use of infantry-borne anti-tank missiles placed well forward was an important feature of the early stages of the Yom Kippur War. It is also important for modern armies to deploy anti-aircraft formations well to the fore. As Napoleon remarked, "it is with guns that war is made". The sight and sound of friendly artillery in action remain important morale considerations, particularly for troops holding defensive posts.

LV

A general should never put his army into cantonments when he has the means of collecting supplies of forage and provisions, and of thus providing for the wants of the soldier in the open field.

ONE great advantage which results from having an army in camp is, that it is easier to direct its spirit and maintain its discipline there. The soldier in cantonments abandons himself to repose. He ends by finding a pleasure in idleness, and in fearing to return to the field. The reverse takes place in a camp. There, a feeling of ennui and a severe discipline make him anxious for the opening of the campaign to interrupt the monotony of service, and relieve it with the chances and variety of war. Besides, an army in camp is much more secure from a surprise than in cantonments, the defect of which usually consists in their occupying too great an extent of ground. When an army is obliged to go into quarters, the Marquis de Feuquières recommends a camp to be selected in front of the line where the troops can be frequently assembled – sometimes suddenly, in order to exercise their vigilance, or for the sole purpose of bringing the different corps together.

CHANDLER: This point could not be more apposite in the modern day in Central Europe. West German formations excepted, NATO forces are stationed in barracks determined by the requirements of the 1945 occupation of a defeated Germany. These do not, unfortunately, match their modern NATO pre-battle positions. Consequently almost all NATO forces have to move to more or less distant alarm posts, which would take up valuable warning time and be likely to attract the attentions in a major crisis of Fifth Column and *Spetznatz* forces. Napoleon was always clear of the perils incurred by an army caught in the process of concentrating forward by an attacking enemy – and demonstrated it at Blücher's expense on 15 and 16 June 1815. On the other hand, the chances of war by mistake are lessened to some degree, at least from the Warsaw Pact point of view, by the fact that NATO forces are in no position to attempt a co-ordinated surprise attack. Napoleon mainly had in mind the advantages of supplementing regular logistic supply by "living off the country" by separating and constantly moving his major formations. Today the complexities of modern forces' armament and supply requirements, and above all the shortages of costly fuel, rule out this ruthless "locust cloud" practice of earlier generations.

LVI

A good general, a well-organised system, good instruction, and severe discipline, aided by effective establishments, will always make fine troops, independently of the cause for which they fight. . . .

THIS remark appears to me less applicable to officers than to soldiers, for as war is not a state of things natural to man, it follows that those who maintain its cause must be governed by some strong excitement. Much enthusiasm and devotedness are required on the part of the troops for the general who commands to induce an army to perform great actions in a war in which it takes no interest. This is sufficiently proved by the apathy of auxiliaries, unless when inspired by the conduct of their chief.

CHANDLER: This maxim is largely self-evident and is as true today as it was in the early 19th century. Discipline and good training (based on regular realistic exercises and manoeuvres) backed by effective administration including logistics are the basic requirements for any effective modern army.

The question of the ideological motivation of the troops is more complex. It may be that the excellent German *Wehrmacht* of 1939–45 served unworthy political masters in the Nazi government, but to some degree a country "gets the army [as well as the government] that it deserves". Thus General Baptista's corrupt regime in Cuba was served by a largely ineffective army and police force, and fell an easy prey to Castro's guerrillas. The questionable Argentinian military *junta* of the late 1970s produced armed forces to match (its air force élite and marines probably excepted). Rotten governments therefore rarely enjoy exceptional armed forces. This said, there is no doubt that patriotic inspiration is of the greatest importance in achieving high morale and thus military effectiveness amongst the young of a nation – the case of modern Israel is very much to the point. At the same time these motivations may be used to conceal fundamental weaknesses and rotteness in the body politic, as in Argentina. But Napoleon was a master at inspirational man-management. "One must speak to the soul; it is the only way to electrify the man."

LVII

When a nation is without establishments and a military system, it is very difficult to organise an army. . . .

THIS is an unanswerable truth, more particularly with reference to an army intended to act upon the system of modern war, and in which order, precision, and rapidity of movement are the principal essentials to success.

CHANDLER: In other words, it is necessary to follow Sir John Moore's maxim: "In time of peace, prepare for war." Proponents of massive unilateral disarmament should take careful heed. One cannot always rely on a period of "phoney-war" (as in 1939–40) to permit the making up of peacetime military deficiencies, whether in whole or in part. This argument can also be deployed in favour of maintaining a system of conscription, thereby ensuring a large reservoir of trained men against an emergency.

LVIII

The first qualification of a soldier is fortitude under fatigue and privation. Courage is only the second. Hardship, poverty, and want are the best school for a soldier.

VALOUR belongs to the young soldier as well as to the veteran, but in the former it is more evanescent. It is only by habits of service, and after several campaigns, that the soldier acquires that moral courage which makes him support the fatigues and privations of war without a murmur. Experience by this time has instructed him to supply his own wants. He is satisfied with what he can procure, because he knows that success is only to be obtained by fortitude and perseverance. Well might Napoleon say, that misery and want were the best school for a soldier, for as nothing could be compared with the total destitution of the army of the Alps when he assumed the command, so nothing could equal the brilliant success which he obtained with this army in his first campaign in Italy. The conquerors of Montenotte, Lodi, Castiglione, Bassano, Arcole, and Rivoli had beheld, only a few months before, whole battalions covered with rags, and deserting for the want of subsistence.

CHANDLER: This plea for hard, realistic training cannot be rebutted. The dangers of "soft" peacetime soldiering have often been revealed at the outset of the next war. The same can be said for lack of proper motivation. The true condition and battle-worthiness of the large and vaunted French army in 1870 and again in 1939 – and the humiliating defeats these inadequate factors led to – are grim enough warnings for modern generations. Long-standing "Alliances for Peace" can become soft as the decades go by without major crises. Both NATO and the Warsaw Pact contain "weaker brethren" who might well prove more of a liability than an asset in a time of major crisis.

LIX

There are five things which a soldier should never be without: his firelock, his ammunition, his knapsack, his provisions (for at least four days), and his entrenching tool. . . .

IT is fortunate that Napoleon has recognised the advantage of giving to every soldier an entrenching tool. His authority is the best answer to the ridicule which has been thrown upon those who proposed it. An axe will be found to inconvenience the foot soldier as little as the sword he wears at his side, and it will be infinitely more useful. When axes are given out to companies, they are soon lost, and it often happens, when a camp is to be formed, that a difficulty arises in cutting wood and building huts for the soldier; whereas by making the axe a part of every man's appointments, he is obliged to have it always with him: and whether the object be to entrench himself in a village, or to erect huts in a camp, the commander of the corps will speedily see the advantage of this innovation.

When once the axe has been generally adopted, we shall perhaps see the desirability of issuing pickaxes and shovels to particular companies, and also the benefit of more frequent entrenchments. It is more particularly during retreats that it is important to entrench when the army has reached a good position, for an entrenched camp not only furnishes the means of rallying troops which are pursued, but if it be fortified in such a manner as to render the issue of an attack doubtful to the enemy, it will not only sustain the *moral* of the soldier in the retreat, but afford the general-in-chief opportunities for resuming the offensive, and profiting by the first false movement on the part of his adversary. It will be recollected how Frederick, in the campaign of 1761, when surrounded by two Russian and Austrian armies, whose united force was quadruple his own, saved his army by entrenching himself in the camp of Buntzalvitz.

CHANDLER: "A soldier's best friend is his rifle" runs a modern barrack-room saw; the sense of this maxim remains true. Possession of proper equipment at all times remains vital for a professional soldier – and weaponry, equipment and uniform need constant up-dating, however costly in financial terms. The need for an improved type of field-service boot in the British Army was one lesson that emerged from the Falklands.

LX

Every means should be taken to attach the soldier to his colours. This is best accomplished by showing consideration and respect to the old soldier. . . .

SOME modern writers have recommended, on the other hand, to limit the period of service, in order to bring the whole youth of a country successively under arms. By this means they purpose to have the levies *en masse*, all ready trained and capable of resisting a war of invasion. But however advantageous at first sight such a military system may appear, I believe it will be found to have many objections.

In the first place, the soldier fatigued with the minutiæ of discipline in a garrison will not feel much inclined to re-enlist after he has received his discharge, more especially since, having served the prescribed time, he will consider himself to have fulfilled all the duties of a citizen to his country. Returning to his friends, he will probably marry, or establish himself in a trade. From that moment his military spirit declines, and he soon becomes ill-adapted to the business of war. On the contrary, the soldier who serves long becomes attached to his regiment as to a new family. He submits to the yoke of discipline, accustoms himself to the privations his situation imposes, and ends by finding his condition agreeable. There are few officers that have seen service who have not discovered the difference between old and young soldiers, with reference to their power of supporting the fatigues of a long campaign, to the determined courage that characterises their attack, or to the ease with which they rally after being broken. Montecuccoli observes that it takes time to discipline an army, more to inure it to war, and still more to constitute veterans. For this reason he recommends that great consideration should be shown to the old soldiers, that they should be carefully provided for, and a large body of them kept always on foot. It seems to me also that it is not enough to increase the pay of the soldier according to his period of service, but that it is highly essential to confer on him some mark of distinction that shall secure to him privileges calculated to encourage him to grow grey under arms, and, above all, to do so with honour.

CHANDLER: All modern armies have progressive pay scales, so veterans (according to rank) derive positive material advantages over the newly joined. The envied and ancient British regimental system (massively adjusted over the years although it has been through merging units and creating large regiments) probably remains the best means of inculcating a high sense of duty and "belonging" in the individual soldier. The United States Army is trying to apply an adapted model (the "cohort-system"), but most armies still retain a "corps of infantry," or "corps of armour" concept, relying on arm or divisional loyalties to foster *esprit de corps*.

LXI

It is not studied speeches at the moment of battle that render soldiers brave. The veteran scarcely listens to them, and the recruit forgets them at the first discharge. . . .

THE opinion of the general-in-chief energetically expressed is, notwithstanding, productive of great effect on the *moral* of the soldier.

In 1703, at the attack of Hornbec, Marshal Villars, seeing the troops advancing without spirit, threw himself at their head: "What," said he, "is it expected that I, a marshal of France, should be the first to escalade when I order *you* to attack?"

These few words rekindled their ardour, officers and soldiers rushed upon the works, and the town was taken almost without loss.

"We have retired far enough to-day; you know I always sleep upon the field of battle!" said Napoleon, as he flew through the ranks at the moment of resuming the offensive at Marengo. These few words sufficed to revive the courage of the soldiers, and to make them forget the fatigues of the day, during which almost every man had been engaged.

CHANDLER: Napoleon was a past-master at propaganda, although not invariably of the wholly truthful variety – the French phrase "*mentir comme un bulletin*" ("to lie like a bulletin") dates from this period. Despite his ruling here, he was in the habit of issuing pre-battle proclamations to his troops – as before Austerlitz in 1805. Sometimes, it is also true, he inserted suitably stirring material into the official record *post facto* for the benefit of the French people and posterity. Thus the 1796 proclamation marking his assumption of command of the Army of Italy is almost certainly a fabrication. Nevertheless the importance of keeping troops informed cannot be over-stressed. Montgomery was very adept at this in Eighth Army and later in North-West Europe; Mountbatten and Slim used the *SEAC Times* and the *14th Army Newspaper* most effectively to counter the "Forgotten Army" complex in Burma, 1943–5.

LXII

Tents are unfavourable to health. The soldier is best when he bivouacks, because he sleeps with his feet to the fire, which speedily dries the ground on which he lies. A few planks or a morsel of straw shelter him from the wind. . . .

THE acknowledged advantage of bivouacking is another reason for adding an entrenching tool to the equipment of the soldier; for with the assistance of the axe and shovel he can hut himself without difficulty. I have seen huts erected with the branches of trees covered with turf, where the soldier was perfectly sheltered from the cold and wet, even in the worst season.

CHANDLER: The modern soldier is trained to bivouac; tentage is rarely resorted to, except at non-tactical administrative base areas. Senior officers today receive command caravans or tracked command vehicles to permit them to set up forward mobile headquarters. These need to be well camouflaged to avoid the attentions of the enemy, particularly in the modern age of airpower. General Geyl von Schweppenberg's Panzer Corps headquarters was effectively wiped out near La Caine in Normandy on 8 June 1944 by an observant Typhoon pilot calling down a highly effective rocket strike from the Allied air forces' "cab-rank" operating in support of the troops ashore in the immediate post-D-Day phase of Operation "Overlord". This attack killed or wounded 19 senior German staff officers on the eve of their launching a massive armoured counter-attack against the Allied beach-head. Its effect was to cause this German operation to be abandoned.

Today's modern sensory and electronic surveillance equipment is so advanced that unless massive jamming and other precautions are taken an opponent can rapidly learn details of his rival's resources. It is no longer a case of counting tents or estimating numbers from visible camp fires. As for the use of houses, modern forward headquarters are often built into them, utilising cellars or other cover. This maxim, therefore, is largely out-dated.

LXIII

All information obtained from prisoners should be received with caution, and estimated at its real value. A soldier seldom sees anything beyond his company; and an officer can afford intelligence of little more than the position and movements of the division to which his regiment belongs. . . .

MONTECUCCOLI wisely observes that prisoners should be interrogated separately in order to ascertain by the agreement in their answers how far they may be endeavouring to mislead you. Generally speaking, the information required from officers who are prisoners should have reference to the strength and resources of the enemy, and sometimes to his localities and position. Frederick recommends that prisoners should be menaced with instant death if they are found attempting to deceive by false reports.

CHANDLER: The acquisition of intelligence has already been referred to at some length. The interrogation of prisoners can still be very valuable for up-dating information and corroboration purposes, but, as the maxim suggests, it requires handling with care. The planting of erroneous information in this way is one of the oldest tricks imaginable. On the evening before Blenheim (1704), for example, Marlborough sent carefully briefed "deserters" into Marshal Tallard's camp with misleading information suggesting that he was in full retreat. In fact he was about to march through the night in seven columns to attack the French early on 13 August. In 1943, Allied intelligence used the corpse of a tramp killed in an air raid to create "the Man who Never Was" – a supposed Royal Marine staff officer carrying details of a non-existent Allied landing operation – in order to conceal the actual forthcoming invasion of Sicily. Today, as already remarked, the problem is profusion of intelligence information rather than its lack. The great difficulty, just as in Napoleon's day, is verification. The intelligence war is a vital component of modern command, communication and control.

LXIV

Nothing is so important in war as an undivided command. For this reason, when war is carried on against a single power there should be only one army, acting upon one base, and conducted by one chief.

SUCCESS, says the Archduke Charles, is only to be obtained by simultaneous efforts, directed upon a given point, sustained with constancy, and executed with decision. It rarely happens that any number of men who desire the same object are perfectly agreed as to the means of attaining it, and if the will of one individual is not allowed to predominate there can be no *ensemble* in the execution of their operations, neither will they attain the end proposed. It is useless to confirm this maxim by examples. History abounds in them.

Prince Eugene and Marlborough would never have been so successful in the campaigns which they directed in concert if a spirit of intrigue and difference of opinion had not constantly disorganised the armies opposed to them.

CHANDLER: "Better one bad general than two good ones" was another of Napoleon's best-known utterances. On the evening of the battle of Lodi in 1796 he learned that the command of the Army of Italy was henceforth to be shared between Kellermann the Elder and himself. He successfully challenged the ruling. He was also adamant in later years that the French Army was a single entity, *la Grande Armée*, from which component parts, the *corps d'armée*, would be temporarily detached to form regional groupings. This was a rejection of the earlier system set up by Lazare Carnot during the French Revolution when more than a dozen armies were brought into existence to defend the frontiers of France and carry the revolutionary message to other countries. This practice had created rivalries between army commanders – including one between Napoleon of the Army of Italy and General Moreau of the Army of the Rhine.

Rigid centralisation of authority was all very well until the creation of a second front, as in the case of Russia in 1812, caused Napoleon to have to try to run the strategic aspects of the perennial Peninsular War in Spain from Dresden, Vilna or even Moscow – with well-known results. But in so doing he was ignoring the phrase "against a *single* power" forming part of this maxim. In modern wars the lesson has been well applied to alliances as well as to national armies. Thus General Eisenhower was Supreme Allied Commander Europe (or SACEUR) during 1944–5 against Hitler's "Thousand Year Reich", while Admiral Mountbatten was Supreme Allied Commander South-East Asia against the Japanese Empire. Today, NATO has a SACEUR (invariably a senior American general); the Warsaw Pact similarly is always headed by a Soviet marshal.

LXV

The same consequences which have uniformly attended long discussions and councils of war will result at all times. They will terminate in the adoption of the worst course, which, in war, is always the most timid; or, if you will, the most prudent. The only true wisdom in a general is determined courage.

PRINCE Eugene used to say that councils of war are only useful when you want an excuse for attempting *nothing*. This was also the opinion of Villars. A general-in-chief should avoid, therefore, assembling a council on occasions of difficulty, and should confine himself to consulting, separately, his most experienced generals, in order to benefit by their advice, while he is governed at the same time in his decision by his own judgment.

By this means he becomes responsible, it is true, for the measures he pursues; but he has the advantage also of acting upon his own conviction, and of being certain that the secret of his operations will not be divulged, as is usually the case where it is discussed by a council of war.

CHANDLER: Napoleon no doubt had in mind here the case of Prussia in 1806, when a hydra-headed command system led to a profusion of plans and growing confusion. Some would say that NATO is not completely free of such a problem in the late 1980s, particularly at the grand strategic or political levels. The regular twice-yearly meetings of the Council of Ministers (and additional emergency consultations) frequently bring out differences of opinion among the member countries – as over the deployment or otherwise of American cruise missiles in Europe or how best to regard the recent Gorbachev offers to withdraw intermediate-range and other missiles. However, a general consensus usually emerges in the end, and a strong Supreme Commander such as the recent General Rogers usually gets his way, although democratic forms of consultation and debate are necessarily gone through. Soviet control of the

Warsaw Pact is more authoritarian; although recently there seem to have been considerable debates among members, the Soviet view invariably prevails.

LXVI

In war the general alone can judge of certain arrangements. It is he alone who can conquer difficulties by his own superior sagacity and courage.

THE officer who obeys, whatever may be the nature or extent of his command, will always stand excused for executing implicitly the orders which have been given to him. This is not the case with the general-in-chief, on whom the safety of the army and the success of the campaign depend. Occupied without intermission in the whole process of observation and reflection, it is easy to conceive that he will acquire by degrees a solidity of judgment which will enable him to see things in a clearer and more enlarged point of view than his inferior generals.

Marshal Villars, in his campaigns, acted almost always in opposition to the advice of his generals, and he was almost always fortunate. So true it is, that a general who feels confident in his talent for command must follow the dictates of his own genius if he wishes to achieve success.

CHANDLER: What Wavell called "the loneliness of high command" requires an effective general to reach and take the hard decisions. This he should not do in isolation – it is clearly his duty to consult his staff and any allied liaison officers – but the decisions are his. As

Napoleon once remarked, "It was not the Roman Army that crossed the Rubicon – it was Cæsar." Or again, "In war it is not the men that matter – it is the man." One thinks of Lord Gort's decision (approved in advance by Churchill) to take the British Expeditionary Force out of the command of the French Superior Commander, General Gamelin, at the height of the disastrous "Battle of France" in late May 1940 in order to head for the Channel port of Dunkirk and there to seek evacuation by the Royal Navy and the "little ships". It was the correct decision, although some French commentators would disagree. General Slim rightly spoke of the "crushing burden of high command", particularly crucial decision-taking under adverse circumstances – for example, his "go or no-go" decision for the Second Chindit Operation when Major-General Orde Wingate advised cancellation because of apparent Japanese preparations (revealed by aerial photographs) to obstruct the glider landing areas. Similarly, Eisenhower had to decide whether or not to postpone D-Day for a month owing to a strong summer gale in the Channel in early June 1944. He backed the judgement of Group-Captain Stagg of the RAF Meteorological Service that there would be a brief respite in the weather over 5 and 6 June and decided: "We go." What Napoleon calls "determined courage" we would today call "moral courage in backing one's own judgement to the hilt".

LXVII

To authorise a general or other officers to lay down their arms in virtue of a particular capitulation, under any other circumstances than when they are composing the garrison of a fortress, affords a dangerous latitude. It is destructive of all military character in a nation to open such a door to the cowardly, the weak, or even to the misdirected brave. . . .

IN the campaign of 1759 Frederick directed General Fink, with eighteen thousand men, upon Maxen, for the purpose of cutting off the Austrian army from the defiles of Bohemia. Surrounded by twice his numbers, Fink capitulated after a sharp action, and fourteen thousand men laid down their arms. This conduct was the more disgraceful because General Winch, who commanded the cavalry, cut his way through the enemy. The whole blame of the surrender fell, therefore, upon Fink, who was tried afterwards by a court-martial, and sentenced to be cashiered and imprisoned for two years.

In the campaign of Italy in 1796 the Austrian general, Provéra, capitulated with two thousand men in the castle of Cosseria. Subsequently, at the battle of La Favorite, the same general capitulated with a corps of six thousand men. I scarcely dare to revert to the shameful defection of General Mack in the capitulation at Ulm in 1805, where thirty thousand Austrians laid down their arms; when we have seen during the wars of the Revolution so many generals open themselves a way by a vigorous effort through the enemy, supported only by a few battalions.

CHANDLER: Napoleon is here denouncing permissive clauses included in orders issued to commanders permitting them to negotiate surrender terms for their commands and themselves under certain clearly defined circumstances. He excepts commanders of fortresses –

rather surprisingly because such officers as Philippon fought on beyond the logical limit at Badajoz in 1812, as did the defender of Bayonne the next year. Under the conditions he indicates the Emperor might have understood the capitulation in later wars of Lieutenant-General Percival in "Fortresss Singapore" (18 February 1942), that of the newly-created Field Marshal von Paulus at Stalingrad (2 February 1943), or of Lieutenant Mills and his garrison of 22 Royal Marines holding the island of South Georgia on 3 April 1982, after a heroic seven-hour battle against an overwhelming Argentinian task force, isolated from all hope of relief and 8000 miles from England.

The whole question of surrendering arms is a very fraught one. Famous examples of defying impossible odds go back to the last stand of the Spartans at the Pass of Thermopylae and take in such determined efforts as Marshal Ney's heroic conduct during the retreat from Moscow (1812) or Davout's defence of Hamburg in 1814 against the overwhelming forces of the Sixth Coalition (which continued for some weeks after Napoleon's first abdication) to the defiance to the death expressed by General Cambronne and the last square of the Old Guard at Waterloo – not to forget Brigadier-General McAuliffe's immortal reply "Nuts!" to a German demand for his surrender in isolated Bastogne in 1944. Japanese troops were notorious for fighting to the bitter end in Burma and the Pacific Island campaigns of the Second World War. Clearly there is no hard-and-fast rule. Japanese fanaticism and General Patton's simplistic views on what constitutes cowardice in the face of the enemy represent differing expressions of one extreme; the attitude of certain Italians holding prepared Libyan frontier defences in the Western Desert (early 1941), the other. In one case they requested the advancing Western Desert Force to save their Italian military honour by firing a few symbolic artillery rounds at their positions, whereupon they hauled up the white flag and came out carrying ready-packed suitcases.

LXVIII

There is no security for any sovereign, for any people, or for any general, if officers are permitted to capitulate in the open field and to lay down their arms, in virtue of conditions favourable to the contracting party, but contrary to the interests of the army at large.
. . .

SOLDIERS, who are almost always ignorant of the designs of their chief, cannot be responsible for his conduct. If he orders them to lay down their arms they must do so; otherwise they fail in that law of discipline which is more essential to an army than thousands of men. It appears to me, therefore, under these circumstances, that the chiefs alone are responsible, and liable to the punishment due to their cowardice. We have no example of soldiers being wanting in their duty in the most desperate situations, where they are commanded by officers of approved resolution.

CHANDLER: Here Napoleon was almost certainly recalling the cases of Marshal Marmont's unauthorised surrender of his forces to the Allies on the outskirts of

Paris in April 1814 and the earlier capitulation of General Dupont's corps, surrounded by superior Spanish forces amidst waterless hills, at Bailen in July 1808. Dupont – a hero in 1805 – was utterly disgraced in late 1808 and incarcerated in a fortress. In 1814 he joined the Bourbons and served as Minister of War on two occasions. In 1944 Marshal Petain – the great hero of Verdun (1916) – was stripped of his baton and imprisoned for life (barely avoiding a death sentence) for his part in signing an armistice with the Germans in late June 1940 and thereafter heading a collaborationist Vichy government. General Galtieri is only the most recent head of state to suffer disgrace and eclipse, in his case for bringing on and then losing a war over the Falklands and for crimes against humanity within Argentina. Peoples can be as remorseless as emperors in punishing failure and surrender. Napoleon never forgave his brother-in-law Murat for joining the Sixth Coalition in 1814 and refused to employ his proffered sabre in the Waterloo campaign.

However, to claim that those who obey a legal order to lay down their arms are as guilty and punishable as the issuer of the actual indefensible instruction is unacceptable. Soldiers cannot shelter behind superior orders where the committal of war crimes are concerned, but that is the only exception to the rule of absolute military obedience recognised today.

LXIX

There is but one honourable method of becoming prisoner of war. That is, by being taken separately, by which is meant by being cut off entirely, and when we can no longer make use of our arms. In this case there can be no conditions, for honour can impose none. We yield to an irresistible necessity. . . .

THERE is always time enough to surrender prisoner of war. This should be deferred, therefore, till the last extremity. And here I may be permitted to cite an example of rare obstinacy in defence, which has been related to me by ocular witnesses. The captain of grenadiers, Dubrenil, of the 37th Regiment of the Line, having been sent on detachment with his company, was stopped on the march by a large party of Cossacks, who surrounded him on every side. Dubrenil formed his little force into square, and endeavoured to gain the skirts of a wood (within a few muskets' shot of the spot where he had been attacked), and reached it with very little loss. But as soon as the grenadiers saw this refuge secured to them they broke and fled, leaving their captain and a few brave men, who were resolved not to abandon him at the mercy of the enemy. In the meantime the fugitives, who had rallied in the depth of the wood, ashamed of having forsaken their leader, came to the resolution of rescuing him from the enemy if a prisoner, or of carrying off his body if he had fallen. With this view they formed once more upon the outskirts, and, opening a passage with their bayonets through the cavalry, penetrated to their captain, who, notwithstanding seventeen wounds, was defending himself still. They immediately surrounded him, and regained the wood with little loss. Such examples are not rare in the wars of the Revolution, and it were desirable to see them collected by some contemporary, that soldiers might learn how much is to be achieved in war by determined energy and sustained resolution.

CHANDLER: This maxim is debatable. Certainly the isolated warrior faced by overwhelming odds is entitled under certain circumstances to surrender – but surely so are small units (such as the Royal Marine detachments in the Falklands and South Georgia) and even large ones (such as those of Singapore in 1941, Arnhem in 1944, or Dien Bien Phu ten years later) when the circumstances make further opposition pointless and risk a large-scale and useless loss of further life to no justifiable purpose. "Irresistible necessity" is capable of varying definitions. Napoleon himself surrendered to Captain Maitland, RN, on HMS *Bellerophon* on 15 July 1815 partly in order to escape from a Bourbon order of arrest which he knew was issued against him. By that time further prevarication was out of the question as his hopes of escaping by sea to the United States had been dashed by the appearance of a British squadron off Rochefort.

LXX

The conduct of a general in a conquered country is full of difficulties. If severe, he irritates and increases the number of his enemies. If lenient, he gives birth to expectations which only render the abuses and vexations, inseparable from war, more conspicuous.

AMONG the Romans, generals were only permitted to arrive at the command of armies after having exercised the different functions of the magistracy. Thus, by a previous knowledge of administration they were prepared to govern the conquered provinces with all that discretion which a newly acquired power, supported by arbitrary force, demands.

In the military institutions of modern times the generals, instructed only in what concerns the operations of strategy and tactics, are obliged to entrust the civil departments of the war to inferior agents, who, without belonging to the army, render all those abuses and vexations inseparable from its operations still more intolerable.

This observation, which I do little more than repeat, seems to me, notwithstanding, deserving of particular attention; for if the leisure of general officers was directed in time of peace to the study of diplomacy, if they were employed in the different embassies which sovereigns send to foreign courts, they would acquire a knowledge of the laws and of the government of those countries in which they may be called hereafter to carry on the war. They would learn also to distinguish those points of interest on which all treaties must be based, which have for their object the advantageous termination of a campaign. By the aid of this information they would obtain certain and positive results, since all the springs of action as well as the machinery of war would be in their hands. We have seen Prince Eugene and Marshal Villars, each fulfilling with equal ability the duties of a general and a negotiator.

When an army which occupies a conquered province observes strict discipline there are few examples of insurrection among the people, unless, indeed, resistance is provoked (as but too often happens) by the exactions of inferior agents employed in the civil administration.

It is to this point, therefore, that the general-in-chief should principally direct his attention, in order that the contributions imposed by the wants of the army may be levied with impartiality, and, above all, that they may be applied to their true object, instead of serving to enrich the collectors, as is ordinarily the case.

CHANDLER: The French occupation of much of Spain, 1808–14, is a case in point. Most of the imposed governors of provinces were severe, and atrocity merely provoked counter-atrocity as the spark of "popular war" was fanned into a flame by the spreading hatred for the invader and by Wellington's supportive presence and operations. Only Marshal Suchet, probably inspired by his wife, the Duchess of Albufera, attempted what we would now call "hearts and minds" policies in Catalonia. These resulted in considerable success, at least for a time. However, they were not typical of French occupation forces elsewhere in the Peninsula, or indeed in Italy, Germany, Poland or Russia.

The knife-edge between counter-productive severity and seemingly exploitable weakness is a tricky one to negotiate -- as the Soviet forces have recently learned over eight years in Afghanistan. Repression only inspires resistance; lax good-nature invites only scorn and insult. The middle-way has to be judged with great care. Napoleon himself was far from adept at getting the balance right: his peaces were always dictated ones. Hitler was even less successful. The most notable cases when conquered countries have been converted into genuine friends and allies in modern times are the *rapprochements* between defeated Austria and Prussia in 1867 (the masterly work of Bismarck) and the conversion of West Germany and Japan into close Allies of the United States after 1945 (the work of Truman, Marshall and MacArthur).

LXXI

Nothing can excuse a general who profits by the knowledge acquired in the service of his country to deliver up her frontier and her towns to foreigners. This is a crime reprobated by every principle of religion, morality, and honour.

AMBITIOUS men, who, listening only to their passions, arm natives of the same land against each other (under the deceitful pretext of the public good) are still more criminal. For, however arbitrary a government, the institutions which have been consolidated by time are always preferable to civil war, and to that anarchy which this last is obliged to create for the justification of its crimes.

To be faithful to his sovereign and to respect the established government are the first principles which ought to distinguish a soldier and a man of honour.

CHANDLER: General Moreau (Napoleon's great rival) took service with the Allies and served with them for some time before being mortally wounded at the battle of Dresden in 1813. Major Quisling's activities in Denmark (1940) merit the same denunciation. No one can ever be justified in betraying his country to an outside enemy, but is it absolutely reprehensible for a senior soldier who totally abhors a regime that has seized power in his

country to use his skills and talents to work for its overthrow, even at the price of employing outside assistance?
Is a "consolidated" government, as the 1901 editor
claims, "always preferable to civil war"? If so, General
de Gaulle might be judged to stand condemned for creating the Free French movement and encouraging the
Resistance against the Vichy regime in France, 1940–4.
However, he was not guilty of handing over his country
to foreigners; rather he worked for France's liberation
from the German occupier and secured its true indepen-
dence in the process. But von Paulus was "turned" in
Soviet captivity after Stalingrad, and emerged in 1945 to
play an important role in setting up the communist state
of East Germany. There clearly has to be a distinction
between true "traitors" and classical "liberators", but
sometimes the dividing line is somewhat blurred.
Americans firmly believe the abhorred General Benedict
Arnold betrayed his compatriots to the British during the
American Revolution. Or did he simply return to his
loyalty as a born subject of the British Crown – a status
he shared with George Washington and other patriots?
Was John Paul Jones a treacherous pirate who used his
knowledge of British ports to surprise Whitehaven in
1778 (near which he had been born and grown-up) with
a landing from the sea (however small), or should he be
regarded as the heroic founder of the United States
Navy? It all depends on one's point of view. But there is
rarely justification for any soldier, however tried, breaking his soldier's oath to his country's accredited government. Yet General Franco did approximately that in
1936, sparking off the Spanish Civil War. In an imperfect

world it would appear that the outcome of such a decision is what – in the last resort – decides an act's morality and validity. Might, alas, still often triumphs over right.

LXXII

A general-in-chief has no right to shelter his mistakes in war under cover of his sovereign, or a minister, when these are both distant from the scene of operation, and must consequently be either ill-informed or wholly ignorant of the actual state of things.

In the campaign of 1697 Prince Eugene caused the courier to be intercepted, who was bringing him orders from the emperor forbidding him to hazard a battle for which everything had been prepared, and which he foresaw would prove decisive. He considered, therefore, that he did his duty in evading the orders of his sovereign, and the victory of Zanta, in which the Turks lost about thirty thousand men and four thousand prisoners, rewarded his audacity. In the meantime, notwithstanding the immense advantages which accrued from this victory to the imperial arms, Eugene was disgraced on his arrival at Vienna.

In 1793, General Hoche having received orders to move upon Trèves with an army harassed by constant marches in a mountainous and difficult country, refused to obey. He observed, with reason, that in order to obtain possession of an unimportant fortress they were exposing his army to inevitable ruin. He caused, therefore, his troops to return into winter quarters, and preferred the preservation of his army, upon which the success of the future campaign depended, to his own safety. Recalled to Paris, he was thrown into a dungeon, which he only quitted on the downfall of Robespierre.

I dare not decide if such examples are to be imitated, but it seems to me highly desirable that a question so new and so important should be discussed by men who are capable of determining its merits.

CHANDLER: The development of radio communications between governments and their commanders in the field has largely vitiated the validity of this maxim. Ever since the American Civil War, governments have been capable of being kept very much in the picture – except possibly in matters of detail (as, for example, initially over the My Lai massacre in Vietnam, but that was a deliberate cover-up).

How far a commander can be expected to refuse orders with which he professionally disagrees is debatable. Some commentators avow that General Wavell was too compliant with Churchill's wishes and allowed himself to be rushed into the expedition to Greece in 1941 and then into a premature reopening of the offensive in the Western Desert (Operations "Brevity" and "Battle-Axe"), without making the Prime Minister sufficiently aware of his misgivings. The consequent failures led to his replacement by Auchinleck. Was von Paulus wrong to try and defend Stalingrad to the last, as ordered by Hitler? And would Napoleon ever have admitted that Ney and the other Marshals were right to refuse his orders at Fontainebleau in April 1814, thus compelling his abdication? One is doubtful. Guderian, to his credit,

twice resigned as Chief of Staff in the face of impossible
orders from Hitler. MacArthur was removed by Truman
and replaced by Ridgeway in 1952 during the Korean
War for fundamental disagreements on how the struggle
should be run. But being removed from command is one
thing; resignation quite another. Examples of blind
obedience to orders given on the spot, "obey now and
argue later" – heedless of the probable outcomes –
include the case of the 69th Regiment at Quatre Bras,
whose colonel was ordered to abandon square formation
and to reform line by the Prince of Orange just before
being hit by French cuirassiers, and in obeying this misguided order lost one of his colours and had his battalion
scattered as a result. Lord Cardigan's acceptance of Lord
Raglan's misconstrued order (conveyed by Captain
Nolan) to charge the Russian guns at the head of the
Light Brigade at Balaclava in 1854 is another example.
In similar fashion, a protesting Kutusov accepted the
Tsar's personal order to attack Napoleon at Austerlitz on
the early morning of 2 December 1805 – despite his great
misgivings – but in this case his sovereign was by his
side.

LXXIII

*The first qualification in a general-in-chief is a cool head; that is,
a head which receives just impressions, and estimates things and
objects at their real value. He must not allow himself to be elated
by good news or depressed by bad. The impressions he receives,
either successively or simultaneously in the course of the day, should
be so classed as to take up only the exact place in his mind which
they deserve to occupy. . . .*

THE first quality in a general-in-chief, says Monte-cuccoli, is a great knowledge of the art of war. This is not intuitive, but the result of experience. A man is not born a commander. He must become one. Not to be anxious; to be always cool; to avoid confusion in his commands; never to change countenance; to give his orders in the midst of battle with as much composure as if he were at ease. These are the proofs of valour in a general.

To encourage the timid, increase the number of the truly brave, to revive the drooping order of the troops in battle, to rally those who are broken, to bring back to the charge those who are repulsed, to find resources in difficulty and success even amid disaster, to be ready at a moment to devote himself, if necessary, to the welfare of the State; these are the actions which acquire for a general distinction and renown.

To this enumeration may be added the talent of dis-criminating character, and of employing every man in the particular post which Nature has qualified him to fill. My principal attention, said Marshal Villars, "was always directed to the study of the younger generals. Such a one I found, by the boldness of his character, fit to lead a column of attack. Another from a disposition naturally cautious, but without being deficient in courage, more perfectly to be relied on for the defence of a country." It is only by a just application of these personal qualities to their respective objects that it is possible to command success in war.

CHANDLER: These comments on qualities required by a good commanding general are as applicable today as at any other period. A cool head, an ability to "see the wood for the trees", the avoidance of becoming obsessed by detail, the ability to pick out the essential from the inessential – these are all clearly *desiderata* in a commander-in-chief. Generals such as Napoleon or (on a lower plain) Slim enjoyed these attributes to a remarkable degree. Others, such as Haig throughout the First World War on the Western Front, saw everything through rose-tinted spectacles: "one more big push" was always to secure final victory. Napoleon's requirements for a "General for all Seasons" sets a very high standard. Wavell called for "firmness of soul" – another phrase for inner conviction and moral courage.

LXXIV

To know the country thoroughly; to be able to conduct a reconnaissance with skill; to superintend the transmission of orders promptly; to lay down the most complicated movements of an army intelligibly, but in a few words and with simplicity; these are the leading qualifications which should distinguish an officer selected for the head of the staff.

FORMERLY the duties of the chiefs of the staff were confined to the necessary preparations for carrying the plan of the campaign, and the operations resolved on by the general-in-chief, into effect. In a battle they were only employed in directing movements and superintending their execution. But in the late wars the officers of the staff were frequently entrusted with the command of a column of attack or of large detachments, when the general-in-chief feared to disclose the secret of his plans by the transmission of orders or instructions. Great advantages have resulted from this innovation, although it was long resisted. By this means the staff have been enabled to perfect their theory by practice, and they have acquired moreover the esteem of the soldiers and junior officers of the line, who are easily led to think lightly of their superiors whom they do not see fighting in the ranks. The generals who have held the arduous situations of chief of the staff during the wars of the Revolution have almost always been employed in the different branches of the profession. Marshal Berthier, who filled so conspicuously this appointment to Napoleon, was distinguished by all the essentials of a general. He possessed calm, and, at the same time, brilliant courage, excellent judgment, and approved experience. He bore arms during half a century, made war in the four quarters of the globe, opened and terminated thirty-two campaigns. In his youth he acquired under the eye of his father, who was an engineer officer, the talent of tracing plans and finishing them with exactness, as well as the preliminary qualifications necessary to form a staff officer. Admitted by the Prince de Lambesq into his regiment of dragoons, he was taught the skilful management of his horse and his sword, accomplishments so important to a soldier. Attached afterwards to the staff of Count Rochambeau, he made his first campaign in America, where he soon began to distinguish himself by his valour, activity, and talents. Having at length obtained superior rank in the staff corps formed by Marshal de Ségur, he visited the camps of the King of Prussia, and discharged the duties of chief of the staff under the Baron de Bezenval.

During nineteen years, consumed in sixteen campaigns, the history of Marshal Berthier was little else but that of the wars of Napoleon; all the details of which he directed, both in the cabinet and the field. A stranger to the intrigues of politics, he laboured with indefatigable activity; seized with promptitude and sagacity upon general views, and gave the necessary orders for attaining them with prudence, perspicuity, and conciseness. Discreet, impenetrable, modest, he was just, exact, and even severe in everything that regarded the service; but he always set an example of vigilance and zeal in his own person, and knew how to maintain discipline, and to cause his authority to be respected by every rank under his orders.

CHANDLER: This description of the requirements in a chief of staff are also irreproachable. These qualities permitted Cadogan to serve Marlborough, Berthier Napoleon, Gneisenau Blücher, or Ludendorff Hindenburg – and even de Guingand to assist Montgomery effectively (if not always earning just recognition in the process). They apply as meaningfully to a chief adviser waging a guerilla-type war as to one involved in conventional operations. Perhaps the most important roles of a chief of staff are to ensure that his commander is kept as fully in the picture as circumstances allow while at the same time to relieve him of as many matters of detail as possible, and to convey his orders crisply, concisely and intelligibly to key subordinates. He should be able to read his master's mind and provide him with loyal, unobtrusive service.

LXXV

A commandant of artillery should understand well the general principles of each branch of the service, since he is called upon to supply arms and ammunition to the different corps of which it is composed. . . .

AFTER having recognised the advantage of entrusting the supply of arms and ammunition for an army to a military body, it appears to me extraordinary that the same regulation does not extend to that of provisions and forage; instead of leaving it in the hands of a separate administration, as is the practice at present.

The civil establishments attached to armies are formed almost always at the commencement of a war, and composed of persons, strangers to those laws of discipline which they are but too much inclined to disregard. These men are little esteemed by the military, because they serve only to enrich themselves without respect to the means. They consider only their private interest in a service whose glory they cannot share, although some portion of its success depends upon their zeal. The disorders and defalcations incident to these establishments would assuredly cease if they were confided to men who had been employed in the army, and who, in return for their labours, were permitted to partake with their fellow-soldiers the triumph of their success.

CHANDLER: Napoleon was, of course, a gunner himself in origin and therefore possibly overstates the case here. Surely it is not only artillery commanders who need to understand the requirements of each branch of the service, but every senior officer, not least the logisticians, as the earlier editor points out. All arms and services must be capable of close integration with one another, not least today of air power in its various manifestations in support of land operations. In the 18th and early 19th century, the Boards of Ordnance and Regiments of Artillery carried out many roles where the provision of munitions, supplies and transport were concerned that have since been allotted to other branches of the service. Obviously the presence of artillery Forward Observation Officers and Air Controllers in the most advanced positions is essential in war to ensure the proper use of available fire- and airpower in support of the other fighting arms.

The proper disposition of the Reserve Artillery park was very important in Napoleon's campaigns from 1805 onwards, just as that of nuclear-capable artillery and missiles would be in any future all-out war on the Central Front. "When it is possible to employ thunder-bolts, their use is to be preferred to that of cannon," wrote Napoleon – but even he could not truly have envisaged the implications of nuclear exchange at either the tactical or strategic level. Possession of such "wea-ponry of the last resort" and reliance upon them to redress unfavourable balances in conventional arma-ments during a total emergency lies at the heart of modern NATO doctrine, and it makes the continuing debate (newly fuelled by the Gorbachev initiatives of spring 1987) about the possibility of reducing super-power nuclear arsenals a matter of the utmost gravity and importance.

Paradoxically, many functions of logistics are once again becoming increasingly civilianised responsibilities in modern armies in this age of the computer. The full implications of this are yet to be experienced.

LXXVI

To reconnoitre accurately defiles and fords of every description. To provide guides that may be depended upon. To interrogate the curé *and postmaster. To establish rapidly a good understanding with the inhabitants. To send out spies. To intercept public and private letters. To translate and analyse their contents. In a word, to be able to answer every question of the general-in-chief when he arrives at the head of the army; these are the qualities which distinguish a good general of advanced posts. . . .*

FORAGING parties composed of small detachments, and which were usually entrusted to young officers, served formerly to make good officers of advanced posts; but now the army is supplied with provisions by regular contributions, it is only in a course of partisan warfare that the necessary experience can be acquired to fill these situations with success.

A chief of partisans is to a certain degree independent of the army. He receives neither pay nor provisions from it, and rarely succour, and is abandoned during the whole campaign to his own resources.

An officer so circumstanced must unite address with courage, and boldness with discretion, if he wishes to collect plunder without measuring the strength of his little corps with superior forces. Always harassed; always surrounded by dangers which it is his business to foresee and surmount, a leader of partisans acquires in a short time an experience in the details of war rarely to be obtained by an officer of the line; because this last is almost always under the guidance of superior authority which directs the whole of his movements, while the talent and genius of the partisan are developed and sustained by a dependence on his own resources.

CHANDLER: This maxim puts its finger firmly on the crucial area of military intelligence. Everything from "Ultra" to "hearts and minds", from undercover duties to interrogation of locals, from the use of satellite reconnaissance to the provision of accurate maps can be read into this paragraph. In the 17th century the equivalent duties were the responsibility of the scoutmaster-general; in the 18th and 19th century they devolved upon the "commander of advanced posts". General Savary served Napoleon in this capacity just as Murray, Scovell and Cocks, together with the leaders of Portuguese partisans and Spanish guerrilla bands, served Wellington well at their respective levels in the Peninsula. Russian partisans under Davydov kept Kutusov informed of every French move in 1812, and today the roles of "irregular" forces including SAS and SBS special units remains equally vital. In the late 19th century and throughout the 20th these roles have been the chief responsibility of Directors of Military Intelligence, from Colonel John Henderson of "Stonewall Jackson" fame during the Boer War, to Brigadier-General Charteris, chief of intelligence to Haig, or Brigadier Thompson, who carried out the same functions for Montgomery in 21st Army Group. Admiral Canaris was head of the *Abwehr* or Army Intelligence to the *Wehrmacht* for most of the Second World War. Today the profusion of electronic and computer-

assisted intelligence sources has transformed the acquisition of intelligence, but the keen, cool appraisal of the
trained intuitive human mind remains in the last
analysis as vital as ever.

LXXVII

Generals-in-chief must be guided by their own experience or their
genius. Tactics, evolutions, the duties and knowledge of an
engineer or artillery officer, may be learned in treatises, but the
science of strategy is only to be acquired by experience, and by
studying the campaigns of all the great captains. . . .

A GREAT captain can only be formed, says the Archduke
Charles, by long experience and intense study; neither is
his own experience enough, for whose life is there
sufficiently fruitful of events to render his knowledge
universal? It is therefore by augmenting his information
from the stock of others, by appreciating justly the dis-
coveries of his predecessors, and by taking for his
standard of comparison those great military exploits, in
connection with their political results, in which the
history of war abounds, that he can alone become a great
commander.

CHANDLER: Most generals learn their trade through
experience of command at every successive level over a
considerable period of years. A few – like Cromwell and
Napoleon – seem to be endowed with supreme military
qualities by Nature or genius. Even the "Great
Captains" should improve their concepts from practice.
One of the most damning things Napoleon ever said was:

"I have fought sixty battles and I have learnt nothing that I did not know at the beginning." Even the brightest of geniuses will become dull and his gifts atrophy unless his skills are continually refurbished and a new edge added to his mental powers – and indeed that proved to be Napoleon's own fate.

The "guiding principles" behind Napoleon's selection of Great Captains remain valid – except possibly, the first. Physical proximity of concentrated forces may be a major liability today for reasons already expounded more than once; rapid concentrations from dispersal followed by equally rapid re-dispersal of forces would seem to be the modern requirement, at least where huge alliance armies are involved.

LXXVIII

Read again and again the campaigns of Hannibal, Cæsar, Gustavus Adolphus, Turenne, Eugene, and Frederick. Model yourself upon them. This is the only means of becoming a great captain, and of acquiring the secret of the art of war. . . .

IT is in order to facilitate this object that I have formed the present collection. It is after reading and meditating upon the history of modern war that I have endeavoured to illustrate by examples how the maxims of a great captain may be most successfully applied to this study. May the end I have had in view be accomplished!

CHANDLER: This is possibly the best known of all Napoleon's maxims. It forms the charter of modern military historians, who have ceaselessly to re-argue and re-demonstrate the value of the study of pre-20th century campaigns, armies and commanders against the glib denizens of "relevance" and modernity, the most extreme of whom avow that nothing that happened before 1945 has any value for the late 20th century officer. This purblind attitude often reveals the paucity of understanding and limited intellectual horizons of such misguided, though often superficially persuasive, champions. None would argue that the study of the past *alone* could properly prepare a modern soldier for the strains of command in a future war. A reader should study the achievements and, above all, the ideas of Zhukov, von Manstein, MacArthur and Giap as well as those of Alexander the Great, Cæsar, Marlborough, Frederick, Napoleon and the rest of the Great Captains. (It is noteworthy that Napoleon excludes Marlborough from his list.) The study of the ancients must also form part of the intellectual diet of the thoughtful soldier who really wishes to master the subtleties of his profession. Weaponry, transportation and communication methods may well have changed out of all recognition over the past 87, nay 40 or even 20 years, and doubtless will continue to do so dauntingly often in an age that is becoming increasingly dominated by the wonders of electronics in their myriad forms – of which America's "Star Wars" projects are only the most recent examples.

However, there are still only seven classic manœuvres of conventional warfare – namely *penetration of the centre*, *envelopment of a single flank*, *envelopment of both flanks*, *attack in oblique order*, *the feigned withdrawal*, *attack from a defensive position* and, Napoleon's favourite gambit, *the indirect approach*. Perhaps the roles of air power and the politico-military complexities of Revolutionary Guerrilla Warfare have posed some novel questions, but the essential problems of waging warfare, of exerting leadership and (at a higher level) generalship under conditions of extreme stress and danger remain unchanged, just as the essentials of human nature – after all, the essential bed-rock or lowest common denominator of all conflicts in whatever age, location or format – are immutable. To ignore the past is to run the risk of falling into the old pitfalls that have been demonstrated time and time again. Charles XII of Sweden should have studied the problems of the Teutonic Knights in their forays into Muscovy; Napoleon should have studied the Swedish débâcle at Pultava; Hitler should have given greater attention to the events of 1812 – above all, the effects of "General Winter", "strategic consumption" and partisan warfare. It is small wonder that Lord Montgomery was moved to write: "In my view one of the basic rules of war is 'don't march on Moscow'."

The correct study of the past – set in its correct political, military and social contexts – has a vital part to play in preparing a future general's mind to understand the present and to permit him to make at least educated guesses at the future. History, after all, "is experience teaching through examples" (Dionysius of Halicarnassus). And, as Santayana declared, "A country [read 'army'] that forgets its past is often doomed to relive it." As it is hoped that this small work has shown yet again to a new generation of soldiers and others interested in the mechanics of warfare (and what thoughtful person dares

not be so in this age of subversionary threats, trans-
national and international terrorism, and above all, of
"unthinkable thoughts" of a possible nuclear holocaust?)
that the *Military Maxims* of the Emperor Napoleon
(1769–1821) have as great a validity as stimuli to further
thought in the late 20th century as ever they did in the
apparently somewhat simpler times of the early 19th. It
may well be that "who dares wins". But no less important is the claim that "who reads learns – and (at least in
some measure) understands". That is the point that
Napoleon is making here, in this last military maxim.

APPENDIX A

THE EIGHT GENERALS MENTIONED IN THE MAXIMS BY NAME

ALEXANDER THE GREAT (356–323BC). The most famous conqueror of the ancient world, he succeeded to the Macedonian throne in 336BC. After campaigning in Greece, he set out to create a vast empire from Asia Minor to Egypt and north-western India, overthrowing the Persian Empire in the process. He died at the age of 32.

CAIUS JULIUS CAESAR (*c.*100–44BC). One of the greatest soldiers of all time, he fought many campaigns in the European and Mediterranean areas and ultimately became dictator of Rome. He was a notable historian of his struggles against Pompey in the civil wars and also of his varied Gallic martial enterprises, as well as a great commander, orator and historian. He was ultimately assassinated on the Ides of March.

PRINCE EUGENE OF SAVOY-CARIGNAN (1663–1736). Refused a commission by King Louis XIV, he transferred his services to the Austrian Emperor in 1683. He rapidly became a famous general against the Turks and later against France as the "Twin Captain" to the Duke of Marlborough. He was also a noted patron of the arts.

MARQUIS DE FEUQUIÈRES (1648–1711). A distinguished French soldier of the reign of King Louis XIV who earned particular renown for his campaigns in the Alps and North Italy. He was a daring cavalry commander who also wrote various treatises on the art of war.

FREDERICK THE GREAT (1712–86). After becoming King of Prussia in 1740, he enlarged his country's terrain at the expense of his neighbours, particularly Austria. He fought several desperate wars against daunting odds but emerged triumphant. A military theoretician as well as an expert practitioner, he wrote several notable works. He was also a musician and patron of the arts.

GUSTAVUS ADOLPHUS (1594–1632). As King of Sweden, the "Lion of the North" is also remembered as the "Father of Modern War". His tactical and administrative reforms revolutionised every aspect of his country's military power – not least his preference for linear formations, cavalry charges *à outrance*, large increase in the numbers and types of guns taken upon campaign, and the setting up of a proper supply system. Above all, he taught the need for professional, long-service armies. After learning his trade fighting the Danes, his intervention in the Thirty Years War led to his greatest success (at Breitenfeld, 1631) and to his death (at Lutzen, 1632). His concepts were widely copied throughout Europe.

HANNIBAL (247–183BC). The greatest Carthaginian soldier, he campaigned against Rome in Spain and, after crossing the Alps with elephants, in Italy, winning numerous battles. He then defended Carthage against the Roman counter-strategy, but was decisively defeated by Scipio Africanus at Zama in 202BC and driven into exile. He committed suicide.

HENRI DE LA TOUR D'AUVERGNE, VICOMTE DE TURENNE (1611–75). King Louis XIV's greatest soldier, he campaigned with great skill and success against the Austrian Empire and the Dutch. A master of mobile warfare, he inspired soldiers such as Marlborough, de Saxe and, in a later generation, Napoleon. He was the first of only three French soldiers to be created marshal-general. His name appears three times in the *Maxims*.

APPENDIX B

THEMATIC TABLE OF MAXIMS

Principles of War	VI, VII, X, XI, XVI, XVIII, XXIII, XXIV, XXV, XXVI, XXVII, XXVIII, XXIX, XXXIII, XXXIV, LV, LXXVII, LXXVII
War of Method	V
War of Marches and Counter Marches	XVII
War of Offensive and Defensive	VII, XIV, XIX
Plan of Campaign	II
Line of Operation	XII, XX
Frontiers	I
Organisation of an Army	LVII
Marches of an Army	II, IV, VI, IX, X, XXI, XXIII
Flank	XXX
Army Corps on Marches	XIII
Command	LXIV, LXVI
General-in-Chief	VIII, XV, XXXI, LXIII, LXV, LXVI, LXX, LXXII, LXXIII, LXXIX, LXXXI, LXXXII, LXXXIII
General-in-Chief of Van and Rear	LXXIV, LXXX
General-in-Chief of Cavalry	LXXXVI
General-in-Chief of Artillery	LXXV
General-in-Chief of Engineers	LXXXV
Chief of Staff	LXXIV
Vanguard	XXXII

FURTHER READING

In the limited compass of this volume it has obviously not been possible to give more than the barest references to most of the historical examples cited. The attention of a reader desiring to delve deeper is drawn to the following short list of useful works (those marked with an asterisk containing useful bibliographies).

Dupuy, Colonel R. Ernest and Colonel Trevor N. Dupuy. *The encyclopedia of military history*. New York and London, 1970; revised editions New York and London, 1976, 1986.*

Eggenberger, David. *A dictionary of battles*. New York and London, 1967.*

Esposito, Brigadier-General Vincent J., and Colonel John Robert Elting. *A military history and atlas of the Napoleonic Wars*. New York and London, 1964, 1980.*

Horward, Donald D. (ed.). *Napoleonic military history: a bibliography*. New York, 1985; London, 1986.*

Laffin, John. *A dictionary of battles*. London, 1986.

Sweetman, John. *A dictionary of battles*. London, 1985.

Young, Brigadier Peter, and Brigadier Michael Calvert. *A dictionary of battles, 1816–1976*. London, 1976.

To supplement William A. Cairnes's references to the Great Boer War in his introduction for the 1901 edition as reproduced above, the reader is directed to:

Packenham, Thomas. *The Boer war*. London and New York, 1979.

Selby, John. *The Boer war*. London and New York, 1969.

Reference may also be made to our modern editor's published works:

Chandler, David G. *The campaigns of Napoleon*. New York, 1966; London, 1967.*

— *Marlborough as military commander*. London and New York, 1973.*

— *The art of war in the age of Marlborough*. London and New York, 1975.*

— *Dictionary of the Napoleonic Wars*. New York and London, 1979.*

— *Atlas of military strategy: the art, theory and practice of war, 1618-1878*. London and New York, 1980.*

— *Waterloo: the hundred days*. London and New York, 1980.*

— (ed.). *A traveller's guide to the battlefields of Europe*. 2 vols., London, 1965.*

— *A dictionary of battle*. London and New York, 1987.*

— *Napoleon's Marshals*. New York and London, 1987.*